Applications of Neural Networks in Robot Control

Contents

About the Editors

Luca Patanè

Luca Patanè is an Associate Professor of Automatic control at the Department of Engineering of the University of Messina, Italy. He received the master's degree in Computer Science Engineering and the Ph.D. degree in Automation and Electronic Engineering from the University of Catania, Italy, in 2001 and 2005, respectively. His research interests include nonlinear systems modelling and control, system identification, soft sensor development, neural networks and learning systems, legged locomotion, insect inspired control systems, and the modelling and control of bio inspired robots. His research activity also addresses bio robotics, autonomous systems, and intelligent control architectures inspired by biological principles. He has authored more than 200 scientific publications, including journal articles, conference papers, and book chapters in the fields of control engineering, artificial intelligence, and robotics. He is a Senior Member of the IEEE and serves as Associate Editor of *ISA Transactions, Frontiers in Neurorobotics,* and *Robotics,* contributing to the editorial activities of leading international journals in automation, artificial intelligence, and robotic systems.

Paolo Arena

Paolo Arena is a Full Professor of Systems Theory and Biorobotics at the Department of Electrical, Electronic and Computer Engineering of the University of Catania, Italy. He received both the degree and the Ph.D. degree in Electrical Engineering from the University of Catania in 1990 and 1994, respectively. His research interests include adaptive and learning systems, biorobotics, nonlinear systems, neural networks, cellular neural networks, collective behaviour in living and artificial systems, and cognitive and bio inspired systems. He has coordinated several national and international research projects, including the European projects SPARK and SPARK II, devoted to biologically inspired robotics and intelligent systems. He has authored more than 300 scientific publications, six books, and several international patents in the fields of systems theory, artificial intelligence, and robotics. He is a Senior Member of the IEEE and has served as Associate Editor of *IEEE Transactions on Circuits and Systems Part I,* contributing to the dissemination and advancement of research in nonlinear systems, neural computation, and intelligent autonomous systems.

Preface

The rapid development of artificial intelligence and machine learning has profoundly transformed the field of robotics. Neural networks have evolved from tools primarily used for perception and pattern recognition into essential components of modern robotic systems, contributing to decision making, motion planning, control, adaptation, and autonomous behavior. The Reprint Applications of Neural Networks in Robot Control was conceived to provide a comprehensive overview of these advances and to document recent developments in this highly dynamic research area.

The motivation for preparing this Reprint arises from the increasing convergence between robotics and artificial intelligence. As robotic systems operate in progressively more complex and uncertain environments, conventional model based approaches are increasingly complemented by data driven methods capable of learning, adaptation, and generalization. At the same time, important challenges remain regarding robustness, safety, interpretability, and real-world deployment. The contributions collected in this Reprint provide valuable insights into these issues and identify promising directions for future research.

This Reprint is addressed to researchers, graduate students, engineers, and practitioners working in robotics, artificial intelligence, control engineering, and related disciplines. It is intended both as a reference for specialists and as an accessible resource for readers seeking an overview of current trends and future opportunities in neural network-based robot control.

Luca Patanè and Paolo Arena
Guest Editors

robotics

MDPI

Editorial

Applications of Neural Networks in Robot Control

Luca Patanè [1,*] and Paolo Arena [2]

[1] Department of Engineering, University of Messina, 98122 Messina, Italy
[2] Department of Electrical, Electronic and Computer Engineering, University of Catania, 95125 Catania, Italy; paolo.arena@unict.it
* Correspondence: lpatane@unime.it

This Editorial introduces the Special Issue "Neural Networks for Robot Control", which gathers contributions that reflect the rapidly growing intersection of machine learning and robotics. The papers collected herein address a wide spectrum of topics—from deep reinforcement learning for locomotion and manipulation to neural network-based perception, trajectory forecasting, and safety-constrained control—spanning wheeled, legged, aerial, and magnetically actuated robotic platforms. Together, they illustrate both the remarkable progress and the persistent challenges at this frontier. This Editorial contextualises these contributions within the broader landscape of the field, identifies the knowledge gaps they help close, and outlines the critical open problems that future research must address to realise robust, efficient, and trustworthy neural robot controllers operating reliably in unstructured real-world conditions.

The use of neural networks in robotics has undergone a qualitative transformation over the past decade. Where early applications were largely confined to supervised perception pipelines or isolated classification tasks, contemporary architectures are now end-to-end protagonists of sensing, planning, and actuation. Advances in deep learning [1], reinforcement learning [2], and hardware accelerators have compressed the gap between learned and model-based control to a degree that would have seemed implausible a decade ago.

Although their benchmark performance in simulation environments is impressive, robust deployment on physical hardware under real-world perturbations remains substantially harder. The challenges are both technical and conceptual. On the technical side, real-time inference on resource-constrained embedded systems, sample efficiency during training, and sensitivity to distribution shift all continue to limit deployment. On the conceptual side, the opacity of large neural networks makes it difficult to provide the safety and reliability guarantees that physical robot operation demands—particularly in human-populated or hazardous settings.

This Special Issue was assembled precisely at this inflection point: the moment when the community possesses powerful tools whose real-world usefulness is bottlenecked not by capability but by robustness, interpretability, and efficiency. The eight contributions collected here address that bottleneck from complementary directions, making this an opportune moment to survey recent developments, identify the gaps the contributions address, and chart the terrain that remains unexplored.

Quadrupedal and legged locomotion is an interesting testbed for neural robot control. Model-based methods have been progressively challenged by reinforcement learning (RL) approaches that sidestep the need for hand-crafted dynamics models and produce controllers capable of generalising across diverse terrains [3,4]. Two broad limitations, however, have remained: (i) the gap between simulation training and physical deployment (the so-called sim-to-real gap), and (ii) the limited interpretability of monolithic end-to-end policies.

Recent work has addressed the sim-to-real gap through domain randomisation and adaptive estimation modules [3], while multi-agent decompositions have shown promise in improving both interpretability and sample efficiency. The contribution by Wang et al. [5] in this Special Issue treats each leg as an independent agent under a multi-agent reinforcement learning (MARL) framework augmented with Riemannian motion policies (RMPs). This decomposition is notable for two reasons: it satisfies the structural conditions required for RMP application to underactuated systems, and it recovers locomotion even with reduced numbers of active legs, demonstrating a graceful degradation that is essential for fault tolerance in the field.

Mechanical design can also facilitate learned control. Scaldaferri et al. [6] investigate end-to-end RL policies co-designed with passive adaptive feet that provide haptic terrain perception. Their results show that proprioceptive enrichment through mechanical compliance measurably improves locomotion robustness, a finding that underscores the importance of the hardware–software co-design paradigm as an alternative to purely computational solutions. Similarly, Al Tawil et al. [7] demonstrate a hybrid locomotion strategy in which passive ventral wheels reduce the cost of transport by a factor of four to five on flat terrain, offering a simple mechanical upgrade with impactful energetic returns.

Classical planning algorithms such as A* offer optimality guarantees but scale poorly with environment complexity. Learning-based heuristics offer a complementary approach: they sacrifice worst-case guarantees in exchange for dramatically lower average-case computational cost. Waga et al. [8] demonstrate this trade-off concretely using an Attention U-Net architecture that expands 62% fewer nodes than A* while achieving near-optimal path lengths. The use of spatial attention mechanisms for path planning, rather than the image segmentation tasks for which U-Net was originally proposed [9], represents a methodologically interesting transfer that merits further exploration in three-dimensional and dynamic environments.

The deployment of neural controllers in safety-critical tasks—aerial vehicles operating near obstacles, firefighting robots, or surgical manipulators—demands formal or empirical guarantees that learned behaviours will not endanger property or life. Control Barrier Functions (CBFs) provide one principled framework for encoding safety constraints [10], but their combination with learning-based controllers has proven technically non-trivial, particularly when the system model is uncertain.

Panja et al. [11] address this directly for quadrotor UAVs performing target-tracking under wind disturbances. Their architecture combines a learned CBF—trained to account for environmental perturbations with a sliding mode outer loop that compensates for unmodelled dynamics. The result is a unified controller that maintains safe leader-following behaviour without relying on an accurate system model. This kind of hybrid solution, where a learned component supplies adaptability and a classical component supplies robustness, reflects a broader trend that the field is increasingly embracing.

Similarly, the neural network-based jet trajectory forecasting system of Pozharkova and Chentsov [12] addresses a scenario—firefighting robots operating under wind disturbance, with partially obscured visibility, where classical ballistic models are unreliable and reactive vision-based feedback is unavailable. The Computational Fluid Dynamics (CFD)-trained network effectively inverts a complex fluid dynamics process in real time, a demonstration of how physics-informed training data can substitute for physical interpretability when direct modelling is intractable.

Robotic manipulation via deep RL has advanced rapidly, yet the sample complexity of policy learning remains prohibitive for many practical deployment contexts. Knowledge Graph Embeddings (KGEs) have been proposed as a mechanism to inject structured semantic priors into otherwise data-hungry learners [13]. Güitta-López et al. [14] validate this

https://doi.org/10.3390/robotics15070117

approach for robotic manipulators, reporting a 60% reduction in learning time and a 15% improvement in task accuracy by combining visual observations with KGE representations. Crucially, these gains are obtained without increasing training time or computational cost, suggesting that semantic structure is a practically deployable technique.

An often-overlooked domain of neural robot control is magnetically actuated micro- and meso-scale robotics, where the impossibility of tethered actuation and the non-linearity of electromagnetic fields make classical control particularly difficult. Huynh et al. [15] present a data-driven regression approach that achieves sub-millimetre positional control of a hard-magnetic sphere within a four-coil array, using a compact neural network trained directly on experimental input–output data. While the system considered is simple, the methodology generalises naturally to more complex arrays and to three-dimensional guidance, making it a foundational result for the emerging field of magnetically steered medical robots.

Reading across these eight contributions, several cross-cutting observations emerge that bear on the future direction of the field.

- **The hardware–software boundary is dissolving.** The clearest signal across contributions is that hardware design and learned control are no longer cleanly separable. Passive wheels, adaptive feet, and custom magnetic arrays are not mere enabling platforms; they are active participants in the control loop. Future progress will increasingly require the co-optimisation of morphology and policy, with differentiable simulators [16] serving as the natural substrate for joint design.
- **Sim-to-real transfer is necessary.** Several contributions rely on simulation for either training or dataset generation. While domain randomisation and physics-informed training data improve transfer fidelity, a theoretical bound on the performance gap between simulation and reality is difficult to define. Addressing this gap—perhaps through conformal prediction [17] applied to robot policies or through online adaptation modules—is a key open problem.
- **Energy efficiency is an underweighted objective.** The hybrid locomotion result of Al Tawil et al. [7] stands out as an explicit engagement with energy cost, yet energetics remain a secondary consideration in most RL and planning work. As robots migrate from laboratory benches to field deployments, the cost of transport and the power budget of onboard inference will become first-class optimisation objectives. Neural architectures designed from the ground up for energy efficiency, including spiking networks [18], deserve greater emphasis.

Drawing together the gaps identified above, we outline five priority research directions that the community should address over the coming years.

- **Neuromorphic and Spiking Neural Network Controllers**. Spiking Neural Networks (SNNs) process information through discrete spike events rather than continuous activations, offering theoretical advantages in energy consumption and temporal processing [19]. Their event-driven nature aligns naturally with sensory data from neuromorphic cameras and tactile sensors [20]. Despite this potential, end-to-end SNN-based robot controllers have remained largely confined to proof-of-concept demonstrations. The integration of SNNs with standard reinforcement learning pipelines, particularly through surrogate gradient methods [21], is maturing rapidly and promises a route to biologically plausible, energy-efficient controllers.
- **Continual and Online Learning**. The papers in this Special Issue train controllers offline and deploy them under a distribution shift. This workflow is appropriate for well-characterised environments but is fundamentally limited when the robot must adapt to genuinely novel conditions. Continual learning methods that update online

https://doi.org/10.3390/robotics15070117

without catastrophic forgetting [22], combined with uncertainty-aware architectures that know when they are operating out of distribution, represent a critical capability gap. Transfer learning from large pre-trained vision–language–action models [23] may provide a valuable solution in this direction.

- **Certified Safety for Learned Controllers**. CBF-based safety filters, as demonstrated by Panja et al. [11], offer one practical path toward certified behaviour. However, the learned CBF itself may be unreliable in regions of state space not covered by training data. Combining conformal prediction with barrier function synthesis [24] could provide distribution-free coverage guarantees without requiring a perfect model. More broadly, the field needs a common evaluation protocol for the safety of neural controllers, identifying the conditions under which safety constraints are violated.
- **Multi-Modal Perception and Sensor Fusion**. The contributions in this Special Issue are largely unimodal: camera images for navigation, joint kinematics for locomotion, or electromagnetic measurements for magnetic actuation. Real-world deployments increasingly demand fusion of heterogeneous sensor streams—vision, depth, proprioception, audio, and tactile sensing—in a computationally tractable manner. Foundation models trained on multimodal data [23] offer one route, but their inference latency and memory footprint are often incompatible with embedded systems. Lightweight multimodal architectures, potentially trained via knowledge distillation from larger models, deserve dedicated investigation.
- **Benchmarking, Reproducibility, and Standardisation**. A pervasive challenge across the contributions collected here is the difficulty of comparing results across platforms, simulators, and experimental protocols. The absence of standardised benchmarks makes it hard to assess whether algorithmic advances are genuine improvements or artefacts of favourable evaluation conditions. Initiatives analogous to ImageNet for perception, or the MuJoCo locomotion suites for RL [25], but covering a broader range of tasks and extending to physical robot experiments, are a community infrastructure priority.

The papers in this Special Issue represent important steps along the underlined path. The Editors express their gratitude to all contributing authors and reviewers whose efforts have made this collection possible.

Conflicts of Interest: The authors declare no conflicts of interest.

References

1. LeCun, Y.; Bengio, Y.; Hinton, G. Deep learning. *Nature* **2015**, *521*, 436–444. [CrossRef] [PubMed]
2. Sutton, R.S.; Barto, A.G. *Reinforcement Learning: An Introduction*, 2nd ed.; MIT Press: Cambridge, MA, USA, 2018.
3. Kumar, A.; Fu, Z.; Pathak, D.; Malik, J. RMA: Rapid motor adaptation for legged robots. In Proceedings of the Robotics: Science and Systems (RSS), Virtual, 12–16 July 2021. [CrossRef]
4. Margolis, G.B.; Yang, G.; Paigwar, K.; Chen, T.; Agrawal, P. Rapid locomotion via reinforcement learning. In Proceedings of the Robotics: Science and Systems (RSS), New York, NY, USA, 27 June–1 July 2022. [CrossRef]
5. Wang, Y.; Sagawa, R.; Yoshiyasu, Y. Learning advanced locomotion for quadrupedal robots: A distributed multi-agent reinforcement learning framework with Riemannian motion policies. *Robotics* **2024**, *13*, 86. [CrossRef]
6. Scaldaferri, A.; Angelini, F.; Garabini, M. Learning to walk with adaptive feet. *Robotics* **2024**, *13*, 113. [CrossRef]
7. Al Tawil, D.O.; Arena, P.; Li Noce, A.; Patanè, L. Quadrupedal locomotion with passive ventral wheels: A data-driven approach to energy efficiency analysis. *Robotics* **2025**, *14*, 158. [CrossRef]
8. Waga, A.; Benhlima, S.; Bekri, A.; Saber, F.Z.; Abdouni, J.; Mzili, T.; Regragui, A. Attention-guided path planning: Learning efficient heuristics for mobile robot navigation via deep neural networks. *Robotics* **2026**, *15*, 96. [CrossRef]
9. Ronneberger, O.; Fischer, P.; Brox, T. U-Net: Convolutional networks for biomedical image segmentation. In *Medical Image Computing and Computer-Assisted Intervention—MICCAI 2015*; Lecture Notes in Computer Science; Springer: Cham, Switzerland, 2015; Volume 9351, pp. 234–241. [CrossRef]

https://doi.org/10.3390/robotics15070117

10. Ames, A.D.; Coogan, S.; Egerstedt, M.; Notomista, G.; Sreenath, K.; Tabuada, P. Control barrier functions: Theory and applications. In Proceedings of the 18th European Control Conference (ECC), Naples, Italy, 25–28 June 2019; pp. 3420–3431. [CrossRef]
11. Panja, P.; Rayguru, M.M.; Baidya, S. Safe autonomous UAV target-tracking under external disturbance, through learned control barrier functions. *Robotics* **2025**, *14*, 108. [CrossRef]
12. Pozharkova, I.; Chentsov, S. Real-time forecasting of a fire-extinguishing agent jet trajectory from a robotic fire monitor under disturbances. *Robotics* **2025**, *14*, 188. [CrossRef]
13. Chen, Z.; Wang, Y.; Zhao, B.; Cheng, J.; Zhao, X.; Duan, Z. Knowledge graph completion: A review. *IEEE Access* **2020**, *8*, 192435–192456. [CrossRef]
14. Güitta-López, L.; Suriani, V.; Boal, J.; López-López, Á.J.; Nardi, D. Boosting deep reinforcement learning with semantic knowledge for robotic manipulators. *Robotics* **2025**, *14*, 86. [CrossRef]
15. Huynh, V.; Mutawak, B.; Do, M.Q.; Ankrah, E.A.; Kassaeiyan, P.; Weinberg, I.N.; Peixoto, N.; Wei, Q.; Mair, L.O. A control interface for autonomous positioning of magnetically actuated spheres using an artificial neural network. *Robotics* **2024**, *13*, 39. [CrossRef]
16. Freeman, C.D.; Frey, E.; Raichuk, A.; Girgin, S.; Mordatch, I.; Bachem, O. Brax—A differentiable physics engine for large scale rigid body simulation. In *Advances in Neural Information Processing Systems 34 (NeurIPS 2021)*; Curran Associates: Red Hook, NY, USA, 2021; pp. 19798–19810.
17. Angelopoulos, A.N.; Bates, S. A gentle introduction to conformal prediction and distribution-free uncertainty quantification. *arXiv* **2021**, arXiv:2107.07511. [CrossRef]
18. Bouvier, M.; Valentian, A.; Mesquida, T.; Rummens, F.; Reyboz, M.; Vianello, E.; Beigne, E. Spiking Neural Networks Hardware Implementations and Challenges: A Survey. *J. Emerg. Technol. Comput. Syst.* **2019**, *22*, 1–35. [CrossRef]
19. Maass, W. Networks of spiking neurons: The third generation of neural network models. *Neural Netw.* **1997**, *10*, 1659–1671. [CrossRef]
20. Gallego, G.; Delbrück, T.; Orchard, G.; Bartolozzi, C.; Taba, B.; Censi, A.; Leutenegger, S.; Davison, A.J.; Conradt, J.; Daniilidis, K.; et al. Event-based vision: A survey. *IEEE Trans. Pattern Anal. Mach. Intell.* **2022**, *44*, 154–180. [CrossRef] [PubMed]
21. Neftci, E.O.; Mostafa, H.; Zenke, F. Surrogate gradient learning in spiking neural networks: Bringing the power of gradient-based optimization to spiking neural networks. *IEEE Signal Process. Mag.* **2019**, *36*, 51–63. [CrossRef]
22. Parisi, G.I.; Kemker, R.; Part, J.L.; Kanan, C.; Wermter, S. Continual lifelong learning with neural networks: A review. *Neural Netw.* **2019**, *113*, 54–71. [CrossRef] [PubMed]
23. Brohan, A.; Brown, N.; Carbajal, J.; Chebotar, Y.; Chen, X.; Choromanski, K.; Ding, T.; Driess, D.; Dubey, A.; Finn, C.; et al. RT-2: Vision-language-action models transfer web knowledge to robotic control. In Proceedings of the 7th Conference on Robot Learning (CoRL 2023), Atlanta, GA, USA, 6–9 November 2023; pp. 2165–2183.
24. Luo, Y.; Ma, Y.; Huang, S.; Zhang, Y.; Schmerling, E.; Pavone, M. Sample-efficient safety assurances using conformal prediction. In *Algorithmic Foundations of Robotics XV*; Springer: Cham, Switzerland, 2023; pp. 149–169. [CrossRef]
25. Brockman, G.; Cheung, V.; Pettersson, L.; Schneider, J.; Schulman, J.; Tang, J.; Zaremba, W. OpenAI Gym. *arXiv* **2016**, arXiv:1606.01540. [CrossRef]

https://doi.org/10.3390/robotics15070117

robotics

MDPI

Article

A Control Interface for Autonomous Positioning of Magnetically Actuated Spheres Using an Artificial Neural Network

Victor Huynh [1,†], Basam Mutawak [1,†], Minh Quan Do [1,†], Elizabeth A. Ankrah [1,2], Pouya Kassaeiyan [1], Irving N. Weinberg [3], Nathalia Peixoto [1,4], Qi Wei [1,*] and Lamar O. Mair [3,5,*]

1 Department of Bioengineering, George Mason University, Fairfax, VA 22030, USA; vhuynh8@gmu.edu (V.H.); bmutawak@gmu.edu (B.M.); mdo9@gmu.edu (M.Q.D.); eankrah@uci.edu (E.A.A.); pkassaei@gmu.edu (P.K.); npeixoto@gmu.edu (N.P.)
2 Department of Informatics, University of California Irvine, Irvine, CA 92617, USA
3 Weinberg Medical Physics, Inc., North Bethesda, MD 20852, USA; inweinberg@gmail.com
4 Department of Electrical and Computer Engineering, George Mason University, Fairfax, VA 22030, USA
5 Image Guided Therapy Research Institute, North Bethesda, MD 20852, USA
* Correspondence: qwei2@gmu.edu (Q.W.); lamar.mair@gmail.com (L.O.M.)
† These authors contributed equally to this work.

Abstract: Electromagnet arrays show significant potential in the untethered guidance of particles, devices, and eventually robots. However, complications in obtaining accurate models of electromagnetic fields pose challenges for precision control. Manipulation often requires the reduced-order modeling of physical systems, which may be computationally complex and may still not account for all possible system dynamics. Additionally, control schemes capable of being applied to electromagnet arrays of any configuration may significantly expand the usefulness of any control approach. In this study, we developed a data-driven approach to the magnetic control of a neodymium magnets (NdFeB magnetic sphere) using a simple, highly constrained magnetic actuation architecture. We developed and compared two regression-based schemes for controlling the NdFeB sphere in the workspace of a four-coil array of electromagnets. We obtained averaged submillimeter positional control (0.85 mm) of a NdFeB hard magnetic sphere in a 2D plane using a controller trained using a single-layer, five-input regression neural network with a single hidden layer.

Keywords: magnetic guidance; neural-network-based controller; untethered manipulation; regression neural networks; machine learning

1. Introduction

The magnetic guidance of particles and devices shows potential for untethered manipulation across a range of medical [1] and industrial applications [2]. Specifically, the manipulation and actuation of particles [3], robots [4], and devices [5] in the milli- to micrometer size range hold particular relevance, as these particles may serve as untethered agents capable of remote-controlled action at a distance [6]. Numerous control algorithms have been developed for the magnetic guidance of various devices [7,8] via an assortment of magnetic coil arrays [9]. Detailed physical modeling approaches have been successfully implemented for controlling magnetic particles [10] and magnetic nanofluids [11] using coil arrays. Often, accomplishing such control requires a reduced-order model of a physical system in which particle motion is driven by external fields [12] (e.g., optical [13], electric [14], acoustic [15], or magnetic [16] fields).

Finite element approaches have been used to model the magnetic fields generated by arrays of coils [17], demonstrating the high-fidelity manipulation of magnetic agents [18]. However, such implementations may be computationally complex and may not innately account for all system dynamics such as coil heating, heterogeneities in the sample environment, or other complexities. Here, we introduced a simple, one-layer, five-input regression neural network for controlling a magnetic sphere in a workspace. Artificial neural networks (ANNs) have a long history in DC motor control [19], with early research applying ANNs for the magnetic guidance of surgical devices such as catheters and implanted magnetic devices [20]. Increasingly, ANNs are being implemented to fine tune magnetic manipulation protocols, gaining ever-increasing precision with iterative training. Prior demonstrations indicate that artificial neural networks can be trained to outperform linear multipole electromagnet modeling relying on fundamental physics, particularly in cases in which iron cores demonstrate nonlinear behavior in the presence of varying applied magnetic fields from electromagnets [21]. Using arrays of electromagnets controlled by ANNs, researchers have positioned millimeter-scale neodymium magnet (NdFeB) disc agents in 2D using an eight-coil array [22], generated real-time predictions of motion dynamics on polymer-based soft magnetic manipulators [23], as well as guided helical microswimmers in 3D [24] and through uncharacterized biomimetic environments [25]. Magnetically guided wheeled robots have been controlled using neuro-fuzzy networks [26], and researchers have made significant progress in guiding endoscopy instruments via intelligent controllers [27–29]. Recently, artificial neural networks have been combined with proportional resonant differential feed-forward control methods for controlling currents in coil arrays aimed at supplying rotational fields to magnetic robots, demonstrating improved control of the robot's position and rotation with extremely small error [30]. Previous work using model-free reinforcement learning aimed at teaching a system to guide a needle using arrays of electromagnets [31] or to suspend a magnetic bead in a fluid against the force of gravity [32,33].

These implementations often require complex, deep, multi-input ANNs requiring significant training via expansive training data sets. Here, we applied a single-layer, five-input ANN architecture to a simple four-coil array, demonstrating the manipulation of a magnetic sphere using magnetic gradients, achieved with no prior training or computational modeling. Using a four-coil array, we achieved the ANN-based manipulation of a hard NdFeB magnetic sphere with average submillimeter precision using a protocol trained on the optical observation of random motions of the sphere in response to random current inputs to a randomly selected coil in the array. We demonstrate that the system is capable of following patterns with both sharp asperities (star-shaped pattern), as well as smooth and continuous curves (circle-shaped pattern). We first describe the mapping of the physical system and delineate the forces involved in actuation. We then describe the experimental apparatus and data collection methodology, including localization and tracking of the NdFeB magnetic sphere. In the sections that follow, we describe the preprocessing of the data and the development of two data-driven controllers (surface fitting model and artificial neural network). We then introduce the ANN developed for manipulating the magnetic sphere and briefly describe system integration and deployment via a custom graphical user interface (GUI). Next, we describe our results, detailing the collected sphere position data set and comparing the accuracy of our ANN approach with the accuracy of a surface fitting model approach. We demonstrate controlled motion along two paths (a star-shaped polygon path and a circular path). Finally, we discuss our control findings and provide some details on the GUI response time, as well as the variability in magnetic sphere detection given different lighting conditions.

2. Methods

In this section, we first establish a system model to provide a basis of the forces involved in magnetic manipulation using our system. Then, we describe our experimental apparatus and data collection protocols, including localization approaches for the electromagnet coils and the NdFeB magnetic sphere. Following that, we describe magnetic sphere tracking, data preprocessing, and the development of two data-driven controllers. Finally, we describe system integration and deployment.

2.1. System Model

Here, we introduce the system with a model that covers the relevant forces involved in magnetic manipulation. We consider a mapping the position of the NdFeB sphere in the workspace as can be described by position $\mathbf{p}$, where $\mathbf{p} \in \mathbb{R}^3$. Magnetic fields in $\mathbb{R}^3$ are generated by currents i (unit, A) running through any electromagnet N_{em} surrounding the workspace [34]. Following an established approach [9], we define the magnetic field $\mathbf{B}$ at a given location $\mathbf{p}$ generated by any single coil carrying current i as $\mathbf{B}(\mathbf{p}, \boldsymbol{i})$, where

$$\mathbf{B}(\mathbf{p},\ \boldsymbol{i}) = \int \mu_0 \frac{i\ \mathbf{dl}\ \times\ (\mathbf{p} - \mathbf{p}_{\mathrm{dl}})}{4\pi\ \|\mathbf{p} - \mathbf{p}_{\mathrm{dl}}\|^3} \tag{1}$$

where $\mathbf{dl}$ is a differential length (unit, m) of the conductor at a specific location $\mathbf{p}_{\mathrm{dl}}$. Assuming N_{em} electromagnets have air cores, we consider the system using a linear magnetic model that follows the principle of superposition, allowing us to write the magnetic field $\mathbf{B}$ for position $\mathbf{p}$ as

$$\mathbf{B}(\mathbf{p}) = \sum_{k=1}^{N_{em}} \mathbf{B}_{\mathbf{k}}(\mathbf{p})\ i_k \tag{2}$$

where i_k is the current (unit, A) supplied to the kth coil. Here, we denote the centroid of the position of the magnetic sphere by $\mathbf{p} = [x, y]^T$. For a NdFeB magnetic sphere having a magnetic moment of $\mathbf{m}_{\mathbf{p}}$, the magnetic field $\mathbf{B}(\mathbf{p})$ applies a force $\mathbf{F}_{\mathbf{m}}$ on the magnetic sphere that is proportional to the magnetic field and field gradient, as given by

$$\mathbf{F}_{\mathbf{m}} = \nabla(\mathbf{m}_{\mathbf{p}} \cdot \mathbf{B}), \tag{3}$$

and all translational motion generated on the magnetic sphere is induced by magnetic field gradients. Accordingly, we denote the linear velocity of the magnetic sphere with

$$\dot{\mathbf{p}} = [\dot{x}, \dot{y}]^T, \tag{4}$$

and the acceleration by

$$\ddot{\mathbf{p}} = [\ddot{x}, \ddot{y}]^T. \tag{5}$$

Here, in addition to the applied magnetic force ($\mathbf{F}_{\mathbf{m}}$), there are additional forces including the the buoyancy force ($\mathbf{F}_{\mathbf{b}}$), gravitational force ($\mathbf{F}_{\mathbf{g}}$), friction force ($\mathbf{F}_{\mathbf{f}}$), and fluidic drag force ($\mathbf{F}_{\mathbf{d}}$). Similarly, magnetic fields induce torques on the magnetic sphere, given by τ_m, where

$$\tau_m = \mathbf{m}_{\mathbf{p}} \times \mathbf{B}, \tag{6}$$

as well as the associated drag torque (τ_{d}), and friction torque (τ_{f}). Numerous works have presented detailed control approaches for the magnetic manipulation of various magnetic agents, and reviews have covered this topic [7,8]. Here, we bypassed all model-based approaches to magnetic agent control and instead developed two regression-based approaches that do not rely on a priori assessments of coils generating magnetic fields in the presence of magnetic objects.

2.2. Experimental Apparatus

Our actuation array consisted of four electromagnets (EMs) positioned in a plane (X–Y plane) and oriented such that coils sat along the cardinal directions (Figure 1a). Opposing EM faces were 40 mm apart, and each EM had an outer diameter of 40 mm, an inner diameter of 22 mm, and a length of 40 mm. Magnets contained 14 layers of windings with 29 turns per layer (AWG 16 magnet wire). We used 3D-printed forms (ULTEM 1010, Stratasys Direct, Rehovot, Israel) to wind the EMs, and iron cores (19 mm diameter, 40 mm length) were placed inside the EMs to increase the fields and gradients generated by the system (Figure 1b).

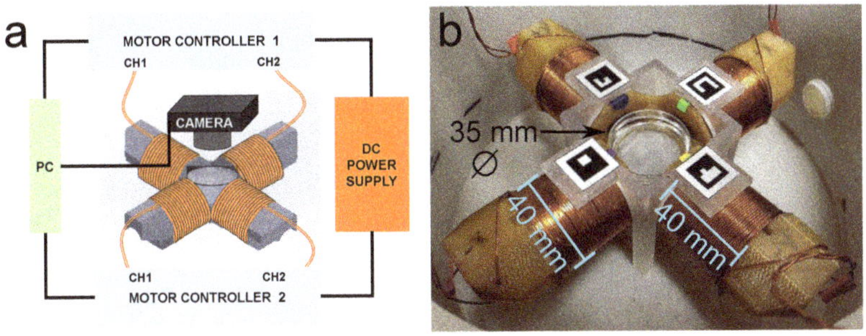

Figure 1. (**a**) Schematic diagram of the experimental setup. (**b**) Picture of the actual setup mounted on a 3D-printed base with labeled dimensions. ArUco markers were assigned to each of the four coils. Centered in the coil array sits a Petri dish (35 mm diameter).

Magnetic fields were generated via PWM signals supplied by two dual-channel motor controllers (RoboClaw, Basicmicro Motion Control, Temecula, CA, USA) (Figure 1a). Motor controllers were powered by 60VDC AC/DC converters (CUI Inc., Tualatin, OR, USA); EMs were controlled by a computer that generated either random current pulses for training or controlled specific current pulses for prescribed planned object manipulation. All training and experimental manipulations were performed in polycarbonate Petri dishes 35 mm in diameter and 10 mm tall (Falcon 351008, Becton Dickinson Labware, Dubai, United Arab Emirates). The Petri dish was filled with hand sanitizer, and the magnetic sphere was placed in the hand sanitizer for manipulation. A camera (Ipevo Ziggi USB Camera, Ipevo Inc., Taipei, Taiwan) was placed above the petri dish to acquire video frames that were streamed to the PC. The video frames recorded the movement of the magnetic sphere over time. Here, we chose a simple 4.8 mm diameter NdFeB magnetic sphere (K&J Magnetics, Pipersville, PA, USA, product # S3, grade N42). Motor controllers were connected to a computer and graphical user interface (GUI) via USB and controlled using packet serial methods based on a C# library. The simultaneous control of the two motor controllers was accomplished via multithreading handled by the GUI.

2.3. Data Collection

An automated data collection process was implemented in which random currents were supplied to the coils sequentially, and the response of the magnetic sphere was recorded. A diagram of the data collection procedure is shown in Figure 2. During data collection, only one coil was ever activated at a given time to produce one data point on the relationship between that coil activation and the resultant magnetic sphere movement. The single-coil actuation constraint significantly simplified the data collection process but also generated an artificially bounded, artificially simple training dataset. Data collection started by determining the position of the magnetic sphere and recording the sphere's starting position (Figure 2A). The protocol then determined the distance between the magnetic sphere and any of the four coils. If the magnetic sphere was within 20 mm of a given coil (Figure 2B), the protocol algorithm automatically selected the opposing coil (Figure 2C). For locations more than 20 mm from any coil, the data collection algorithm selected a coil at random for actuation (Figure 2D). The selected coil was activated with a random current scale value, which was recorded and collated with the resulting sphere motion (Figure 2E,F). The above steps were then repeated until stopped. The recorded data were saved in a CSV file for easy examination and analysis.

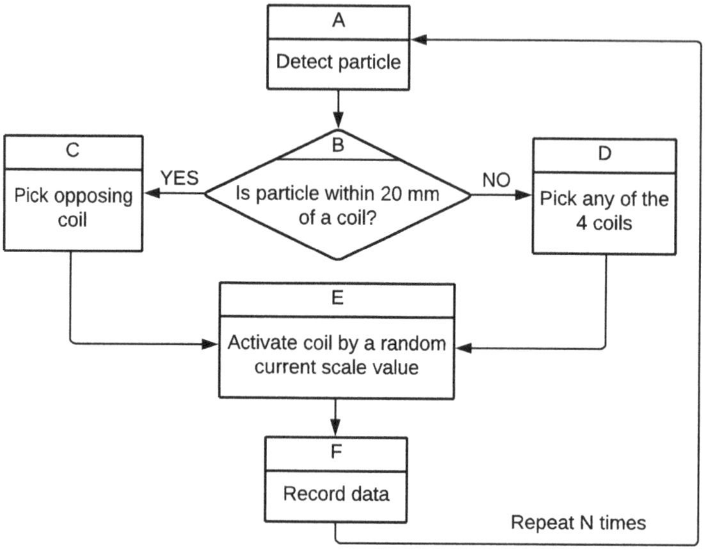

Figure 2. Data collection procedure flowchart.

2.3.1. Coil Localization

The object detection module was also tasked with orienting and calibrating the positions and poses of all EM coils relative to the workspace. To accommodate >30 frames per second imaging and operation, localization of the EM coils was performed in <190 ms. EM coil detection was performed via recognition of coil-specific ArUco fiducial markers (OpenCV) placed on top of each coil [35] (see Figure 1b). The method applied an adaptive threshold to obtain black borders of each ArUco marker, then contoured the image for candidate markers. Following contouring, the object detection module rejected candidates based on an internal filter. The binary pattern of each remaining contour was then analyzed, and the four corners of a marker (defining its location) were returned. For simplicity, here, we refer to individual coils as +X, −X, +Y, and −Y coils.

2.3.2. Magnetic Sphere Localization

An object detection module was created for receiving and processing the optical data collected via the camera, including detection of the magnetic sphere across streamed frames. For magnetic sphere localization in each frame, we first obtained a preoperative comparison image (background image), which could then be subtracted from subsequent streamed frames. Following subtraction, images were segmented within the area of interest (Petri dish region of the image) and then filtered to obtain the magnetic sphere's location. The background image was generated by taking and averaging sequential images of the system with the magnetic sphere positioned at the face of each EM coil via the application of a magnetic field gradient pulling the sphere toward the coil. By actuating each EM sequentially, we positioned the magnetic sphere at the face of each EM coil, collecting an image at each position (Figure 3a–d). We repeated the process, collecting a total of two example images for each of the four magnetic sphere positions. The eight images were then averaged to mitigate the presence of a magnetic sphere, and this averaging step generated a background image that was sufficiently clean to allow for system tracking in future images in which the magnetic sphere could be positioned anywhere in the workspace. Figure (Figure 3e) depicts the generated background image created from the above process of averaging eight images having the magnetic sphere in front of each EM coil. Figure (Figure 3f) presents an image of the Petri dish with the magnetic sphere completely removed from the dish. Our goal with image processing was to accomplish magnetic sphere localization in less than 35 ms, as longer delays in sphere localization result in delays in camera frame acquisition.

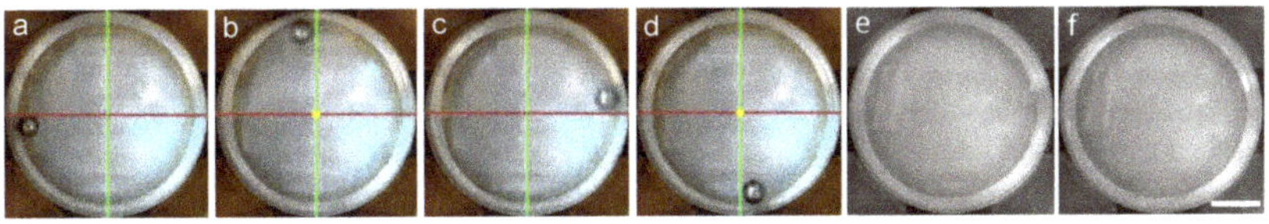

Figure 3. Background image generated by image averaging works well. (**a–d**) Images of a magnetic sphere positioned in front of EM coils. (**e**) Background image created by averaging (**a–d**). (**f**) Background image was taken with the magnetic sphere removed from the Petri dish. The scale bar is 1 cm.

2.3.3. Magnetic Sphere Tracking

A user can interactively draw the path for the sphere to follow using the GUI.The system can control the magnetic sphere to follow any trajectory, including those with sharp turns. The system has two customizable parameters for path configuration. The first one is the path interpolation amount, which determines the discretization resolution of the given path. Even if the specified path has only a few anchor points, with a small interpolation amount the path can be finely discretized so that the magnetic sphere can continuously follow a smooth and interpolated path. The second parameter is the path tolerance, which specifies the acceptable distance of the sphere from a path point within which the sphere is considered as successfully following the corresponding path point, and the next target point is then updated. With these two parameters specified by the user, our system ensures locally smooth path specification and sufficient tracking resolution so that any target path can be followed.

2.4. *Preprocessing*

The collected data were preprocessed before use in model training. Our preprocessing procedure involved first removing all data samples where current scale values were zero. This means that the +X coil input only included values for which the +X coil current was not zero; the same applied for the other coils. Since coils were activated one at a time, the original dataset contained a large proportion of instances in which the current for a given coil was zero, which should be excluded in model training. Following the initial cleaning, approximately 15% of the dataset ($\approx$4500 instances) remained for the training set.

Once data were cleaned of zero values, position data were normalized to a common spatial scale (2D, X–Y plane). Initial x and y positions of the sphere, X_0 and Y_0, and final positions, X_f and Y_f, were all in the range of -17.5 mm and 17.5 mm. The measured distance of the sphere to a coil D was in the range of 5 to 40 mm. Z-score normalization was applied to all data.

2.5. *Development of Two Data-Driven Controllers*

Two data-driven controllers were independently developed using the same collected data for comparison. The first controller is an ANN-based controller, which learns a predictive model to determine the optimal control signal to move the magnetic particle in the desired direction by the desired travel distance. The second controller is a surface-fitting model, which results in polynomial functions of the distance between the particle and a coil and the desired travel distance. Both models learn the relationship between coil current levels and particle location and travel distance but in different ways. Figure 4 depicts a block diagram of the magnet control system with two controllers. The controllers are detailed in the following sections.

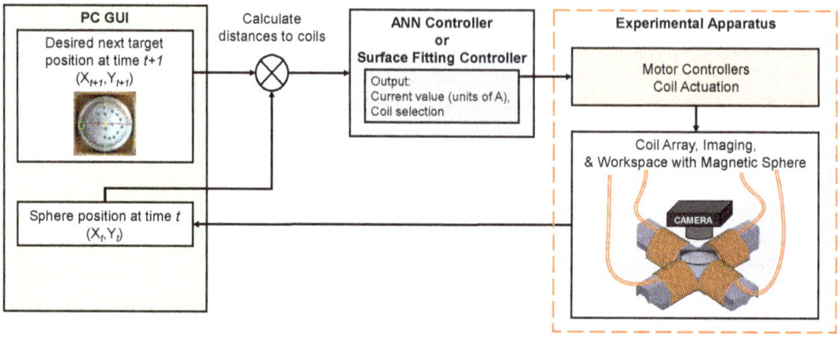

Figure 4. Block diagram of the magnet control system.

2.5.1. Artificial-Neural-Network-Based Controller

Artificial neural networks were used to control sphere movement in the 2D plane. Four ANNs of the same architecture were trained for the four coils that moved the sphere in the x and y directions (Figure 5). To choose the architecture, the number of hidden layers was varied between 1 and 3, and the number of hidden neurons was tested at 10, 15, 20, 30, and 60 systematically. One hidden layer and ten sigmoid logistic neurons were determined as the best architecture parameters. More complex architectures did not provide any significant increase in performance (measured using the R^2 correlation coefficient). Here, we used ANNs based on the generalized regression neural network approach [36]. Each ANN took in five input features, X_0, Y_0, X_f, Y_f, and D as defined above. These five inputs were chosen by applying Scikit–Learn's mutual information regression feature selection method [37] to determine the most relevant features for inclusion. All inputs are expressed in millimeters. Each ANN contained one hidden layer with ten hidden nodes. The output of the ANN was the predicted current scale value of the corresponding coil that would move the sphere to the destination.

Input data were the five measurements collected from tracking the motion of the spherical magnetic sphere as a result of randomly generated pulses applied to the coils. The collected data were split in such a way that 80% was used for training and 20% for testing. Training and testing the ANN models were implemented in the Matlab Deep Learning Toolbox [38] using the Levenberg–Marquardt algorithm [39]. Training the ANN model does not require a specialized computer workstation but can be performeds on any laptop or desktop that has Matlab installed. In our study, training the ANN model took about 10 min only.

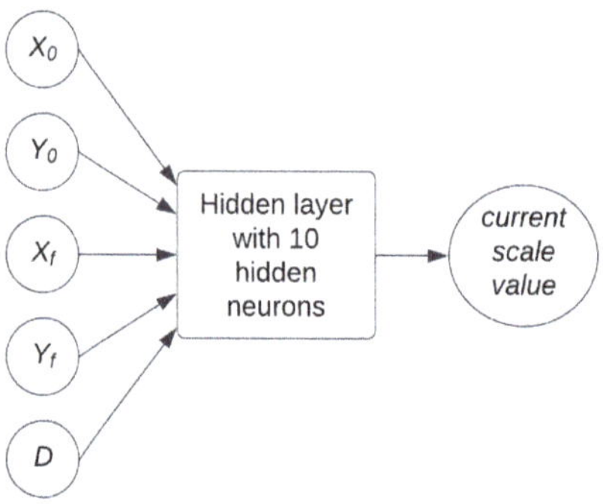

Figure 5. Four ANNs of the same architecture were trained to compute the current scale values of the four coils. X_0, Y_0: initial x and y positions of the sphere, respectively; X_f and Y_f: final x and y positions of the sphere, respectively; D: measured distance of the sphere to a coil.

2.5.2. Surface Fitting Model-Based Controller

The other control method that we implemented for comparison was the surface fitting model. The same collected data described previously werew used to learn a direct mapping between the applied current scale to a coil and the resultant sphere position. System states such as the viscosity of the liquid agent that the sphere traveled in and hardware setup were also captured by the mapping. A nonlinear regression model was used to represent the mapping, which could then be used to predict the amount of coil current needed to move the sphere in the desired direction by the desired distance. Four models, one for each coil, were implemented by using the 'polyfit' function in Matlab. Figure 6 illustrates one fitted surface model, which represents the nonlinear association between the sphere distance from a coil and the traveled distance to the current scale.

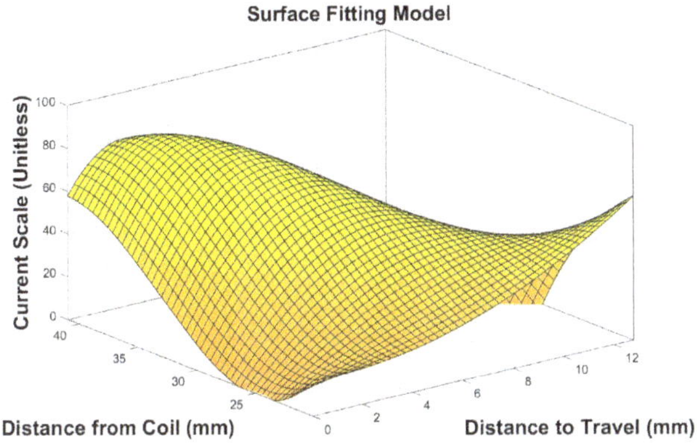

Figure 6. An example of the learned surface model, which maps the sphere distance to the corresponding coil and the traveled distance of the sphere to the current scale applied.

2.6. System Integration and Deployment

Each trained neural network model's weights and biases were exported to our C++ program so that they could be used with the program controlling the actual system. The C++ matrix library Eigen [40] was used to initialize the weight and bias values for each network. User interaction with the system followed a sequence of steps, starting with system calibration, followed by session setup, after which the user drew a path within the Petri dish area and initiated the controller to move the sphere along the drawn path (Figure 7).

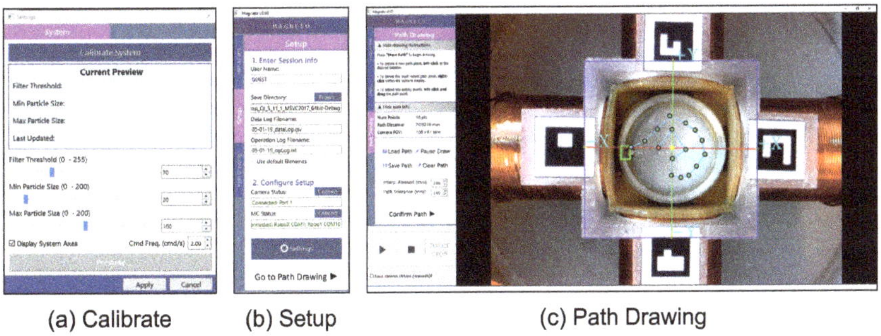

Figure 7. GUI allows the user to first (**a**) calibrate the system, choosing filter thresholds, and minimum and maximum particle sizes (pixels). (**b**) Setup and user–directed file naming, camera status, and motor controller status. (**c**) How a user interacts with the system. First, the start location, end location, and waypoints are selected, and system control is initiated.

3. Results

3.1. Data Collection

Training data collection included a total of 30,000 instances collected over 3 days. Of the 30,000 instances, approximately 19,000 instances were deemed usable for the training protocol. Unusable instances were characterized by insufficient magnetic sphere motion, unclear visualization of the magnetic sphere, or poor magnetic sphere localization. The configuration of the four coils and the constraint on actuating a single coil at a time made it challenging for the magnetic sphere to move to cover the entire workspace. As such, a majority of the observable instances were constrained to the central and +-shaped central region of the workspace. Figure 8 shows the trajectories of the magnetic sphere over the training period. The majority of the sphere movement was along the horizontal and vertical axes. Much less data were collected in "dead zones", depicted by regions with a lack of data points (red + signs with green circles).

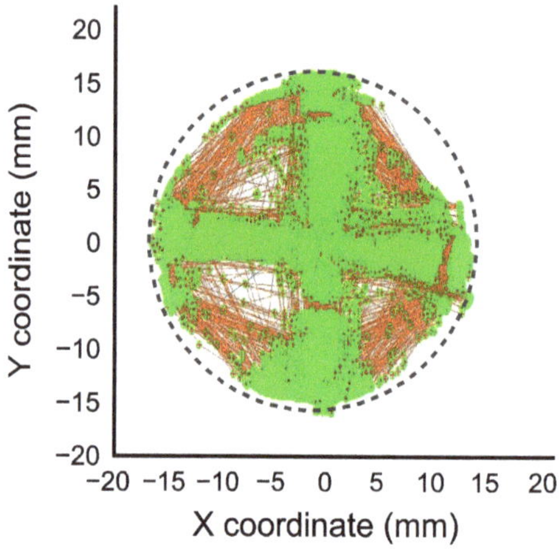

Figure 8. Training data were collected continuously by tracking magnetic sphere positions after the application of a sequence of randomly distributed magnetic pulses. Pulse protocol as described above.

3.2. Accuracy Comparison of Predicted Current Scales from Two Models

Control current scales predicted by the ANN model and surface fitting models were compared to the applied current scales in the test data. R^2 values were calculated to assess the accuracy of the predicted control currents of each model. Figure 9a shows the current scales predicted by the ANN model in comparison to the expected current scales of all the test data instances. The R^2 value of the +X coil instances was 0.93 and that of the −X coil was 0.88, both showing the efficacy of the ANN in generating the control currents to move the sphere by the desired distances. In particular, the predicted model showed a close association with the expected current value when they were under a scale factor of 50. Figure 9b shows the current scales predicted by the surface fitting model in comparison to the expected current scales of the test data. The R^2 values of the +X and −X coils were 0.65 and 0.73, respectively. A comparison of the R^2 values of both horizontal coils showed that the ANN controller performed significantly better than the surface fitting model, with an increase in the R^2 values of the current scales of 0.28 and 0.15, respectively, as predicted by the ANN controller.

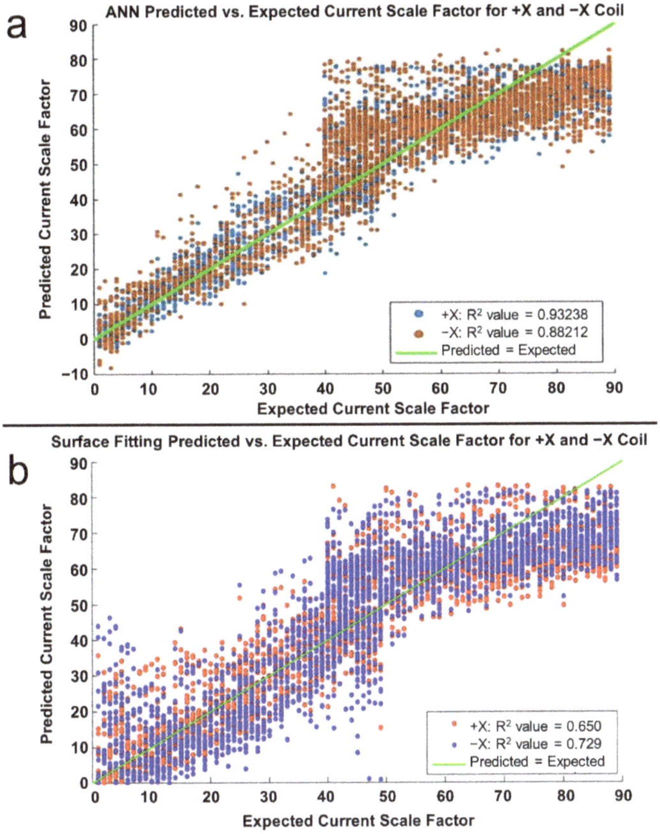

Figure 9. Performance comparison of current scale factor inference between the ANN +X and −X coil models (**a**) and the surface-fitting +X and −X coil models (**b**).

Table 1 compares the performance of all coils controlled by the ANN model to those controlled by the surface fitting model. The ANN controller outperformed the surface fitting model for all coils, yielding R^2 values at least 10% larger than those obtained with the surface fitting model. The averaged ANN model R^2 value was 0.91, while that of the surface fitting model was 0.735, demonstrating the significantly better accuracy of the ANN controller.

Table 1. R^2 values for each coil based on the surface fitting model and the single-layer, 10-neuron artificial neural network.

Trained Coil	Surface Fitting R^2	Artificial Neural Network R^2
+X	0.650	0.932
−X	0.729	0.882
+Y	0.735	0.887
−Y	0.827	0.934
AVERAGE	0.735	0.910

3.3. Comparison of Magnetic Sphere following Trajectories

Both ANN and surface fitting translation models were tested using the same starting sphere location and proposed pathway. Both models allowed translation toward the desired final location following the given path. However, as shown in Figure 10a, the surface fitting model induced frequent and dramatic (several sphere body lengths) overshoot of the sphere position at locations near the coils, especially near the +X and −Y coils. On the other hand, the sphere path that was controlled by the ANN-based controller (shown in blue) was significantly more even throughout the workspace and more closely followed the desired path than the path from the surface fitting model (shown in green). The results suggested that the ANN controller outperformed the surface fitting controller in moving the magnetic sphere. After training, the GMU-Magneto system was capable of controlling the magnetic sphere (4.8 mm diameter NdFeB sphere) with submillimeter precision, using a trained ANN having a single hidden layer with 10 nodes. Across all trials, the average standard deviation from the path for the magnetic sphere was 0.85 mm. The reported results demonstrated the overall bound on the stability of the system. The four actuators were activated one at a time in a sequence. If any single actuator was unreliable, it would have contributed to errors in the overall performance.

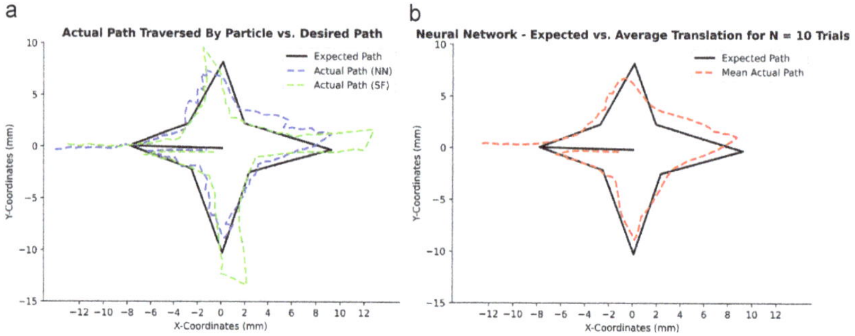

Figure 10. Path traversal comparison. Desired and actual paths traversed by the particle using the two different motion models (**a**). The average path traversed by the particle across N = 10 trials using the neural network model (**b**).

4. Discussion

Here, we developed a system that provides a test bed for examining the application of artificial-neural-network-based controllers in object manipulation using electromagnetic fields. Here, we made use of simple and sparse training data in which the overall sampling of the workspace was relatively incomplete. Even though Figure 8 contains a significant amount of the unsampled space, the controller performed well across the entire workspace, as evidenced in Figure 11. A video of the controller moving the magnetic sphere is included as Supplementary Material. ANN controllers were demonstrated to have reasonable (submillimeter) positional resolution in a 2D setting when trained using our simple approach. An additional limitation in the training data set was that the supplied coil current was binary, either 0 A or 20 A. This means that smaller incremental motions of the sphere, as would be expected for coil currents between 0 A and 20 A, were not represented in the training dataset. One future direction is to design a better training process so that the training data incorporate a richer set of positions and trajectories. We anticipate significant improvements in the system with the addition of incremental coil currents. Significantly, the ANN used here contained only one hidden layer, suggesting that fairly simple networks are capable of positional control of simple magnetic objects with accuracy being significantly smaller than the object diameter. More complex workspaces, for example, workspaces having surfaces with varying coefficients of friction, may be ideal testbeds for future experimentation, as computational assessment of varying friction over a workspace is likely challenging to model well.

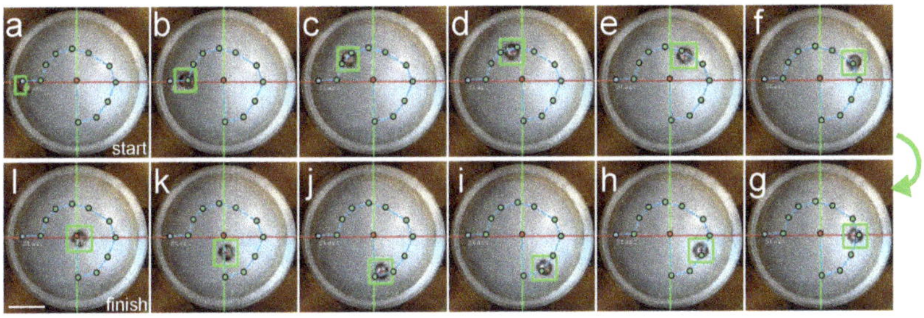

Figure 11. Sphere manipulation demonstration. Using a GUI, a user-defined pattern is proposed having a start position (**a**), 10 intermediate waypoints (**b–k**), and a finish position (**l**). Despite the overwhelming data on sphere response to the random distribution of pulses being centered on the cardinal axes of the workspace (Figure 8), the system readily moves the magnetic sphere through a broad range of the workspace (traversing 270° around the center point, (**a–j**)) and finally returns the magnetic sphere to the center of the workspace (**k**,**l**).

The system presented here has implications for quickly establishing the ANN-based control of magnet objects via electromagnets. For example, the method we presented for learning control of a spherical magnetic sphere may, in the future, be extended to teach coil arrays to control broad libraries of spheres with diverse magnetic properties. Additionally, the presented method may be applied to highly asymmetric arrays of electromagnets to expedite and simplify expectations of how current supplied to such coils would move a magnetic sphere of a given, and perhaps unknown, magnetic ordering. Additionally, expanding on our approach by applying it to systems of moving electromagnets or permanent magnets may allow for control approaches for complex magnet array systems having numerous variables (magnet position, coil current, magnet angle) to be explored quickly, in a randomized fashion, for the creation of a reliable, low-error control approaches. Future applications of the method may be implemented in other fields using magnetic manipulation, such as microfluidics [41], magnetic drug targeting [42], or the future of microscale devices for medicine [43,44]. Additionally, comparing our ANN approach with control methods based on analytical models may be a future method of benchmarking the ANN results and determining how the ANN may be improved in the future. Combining such control efforts with tracking of more complex and medically relevant devices such as surgical needles [45] and incorporating other nonoptical imaging modalities such as ultrasound (US), computed tomography (CT), or magnetic resonance imaging (MRI) may expand the toolbox of magnetic surgical instrument guidance. However, the overall manipulation of surgical instruments requires significant on-the-fly tunability, and control paradigms that can begin to address the dynamics of in vivo operation are only just now emerging [46,47].

4.1. Magnetic Sphere Detection Sensitivity Analysis

To assess system tolerance to various lighting conditions, we recorded the sphere location under three different lighting scenarios: dark room of ≈6 lux (Figure 12a), ambient overhead room lighting of ≈80 lux (Figure 12b), and direct overhead lighting supplied by a lamp of ≈220 lux (Figure 12c). For each determination, a video of the sphere was collected for 60 s, during which time the sphere was not manipulated. During the 60 s video under each lighting condition, 50 frames were collected and segmented to estimate sphere location. The standard deviation in sphere location for each lighting condition was computed. Dark room, ambient, and direct overhead lamp lighting conditions yielded standard deviations in sphere position of 0.34 mm, 0.05 mm, and 0.13 mm, respectively. We suspect that the ambient lighting condition yielded the lower standard deviation in sphere position because the dark room condition provided insufficient illuminance to reliably determine the edges of the sphere, and the direct overhead lamp condition yielded saturated pixels and shadows that also inhibited accurate edge detection in our object detection scheme.

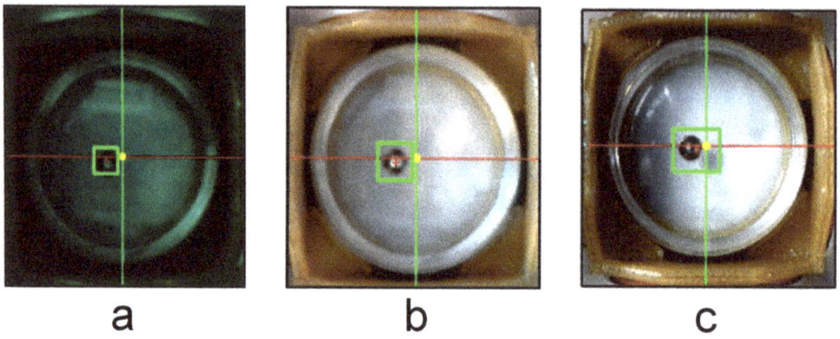

Figure 12. Comparison of sphere location detection in various laboratory lighting scenarios. (**a**) Dark room with only sunlight, approximately 6 lux. (**b**) Ambient lighting from room lights, approximately 80 lux. (**c**) Direct lighting from lamp positioned over Petri dish and coil array, approximately 220 lux.

4.2. GUI Response Time

For the GUI to be practical, reasonable response times are required for a seamless user interface. We quantified system response times to ensure smooth operation, as shown in Table 2. Specifically, we assessed the GUI functions of the startup, camera connection, motor controller connection, opening the settings window, applying settings window edits, starting/pausing system operation, and stopping hardware execution. We found that startup and camera connection initiated long delays, with each operation taking $\approx$1 s. Beyond those two functions, all other functions were completed in significantly less than 100 ms.

Table 2. Typical lag time values for significant GUI functions

Function	Average Lag Time (ms)	Standard Deviation (ms)
Startup	1287.13	16.59
Camera Connection	987.38	182.76
Motor Controller Connection	10.81	0.59
Open Settings Window	52.38	2.11
Apply Settings Window Edits	0.16	0.10
Start/Pause System Operation	1.98	0.56
Stop Hardware Execution	0.09	0.08

Future work demonstrating operability in an array of light sources, and possibly under actively changing lighting conditions, may shed light on how the anticipation of dynamic lighting conditions could be incorporated into the training set for untethered magnetic manipulation systems. Various laboratories have applied advanced learning techniques such as reinforcement learning to the problem of controlling small magnetic devices [48–50], and we acknowledge that the application of such learning approaches would improve the performance of our controller. However, here, we emphasize the ability to control a simple magnetic sphere using very simple regression models and a small sample data set. Additionally, future efforts may include incorporating finite element model simulations for the controller and coil array. Finally, the proposed ANN model can be improved by adding the desired travel time as another input to provide transient movement control in addition to the current position control.

5. Conclusions and Future Work

We developed a simple, easy-to-implement AI-based controller for an electromagnetic system, tested its performance in a four-coil system, and compared it to a surface fitting controller. We described the electromagnet system, magnetic sphere tracking method, our approach for data collection, and ANN structure. We then quantified the ANN's performance in comparison to that of a surface fitting control approach. Our results showed that a simple one-layer artificial neural network could be conveniently trained to control the magnetic sphere's position to follow desired trajectories. The significance of the method presented lies in its simplicity, in that no prior physics knowledge is used in the ANN, and the collected data rely on the actuation of a single coil at a time, with only a single current value used in the data collection. No assumption needs to be made on the system configuration or characteristics, as the ANN model was able to learn them through the training data. Naturally, more complex exploration of the parameter space, such as activating multiple coils at a time or activating coils with incremental current values (i.e., currents between 1 A and 20 A) would significantly expand the original data set and improve the accuracy and precision of the ANN-based controller.

Supplementary Materials: The following supporting information can be downloaded at: https://www.mdpi.com/article/10.3390/robotics13030039/s1, Video S1: Operation of the GUI and positioning of the magnetic sphere.

Author Contributions: Conceptualization, V.H., B.M., M.Q.D., E.A.A., I.N.W., L.O.M. and Q.W.; methodology, V.H., B.M., M.Q.D., E.A.A., P.K., L.O.M. and Q.W.; software, V.H., B.M., M.Q.D.; validation, V.H., B.M., M.Q.D. and E.A.A.; formal analysis, V.H., B.M., M.Q.D., E.A.A., P.K. and Q.W.; investigation, V.H., B.M., M.Q.D. and E.A.A.; resources, I.N.W., L.O.M. and Q.W.; data curation, V.H., B.M., M.Q.D. and E.A.A.; writing—original draft preparation, V.H., B.M., M.Q.D., E.A.A., L.O.M. and Q.W.; writing—review and editing, V.H., B.M., M.Q.D., E.A.A., P.K., I.N.W., N.P., L.O.M. and Q.W.; visualization, V.H., B.M., M.Q.D. and E.A.A.; supervision, I.N.W., N.P., L.O.M. and Q.W.; project administration, I.N.W., L.O.M. and Q.W.; funding acquisition, I.N.W., L.O.M. and Q.W. All authors have read and agreed to the published version of the manuscript.

Funding: This research received no external funding.

Data Availability Statement: Software for the system is available at the following link: https://github.com/bmutawak/MagnetoSuture, accessed on 10 January 2024.

Conflicts of Interest: Authors Irving N. Weinberg and Lamar O. Mair were employed by the company Weinberg Medical Physics, Inc. The remaining authors declare that the research was conducted in the absence of any commercial or financial relationships that could be construed as a potential conflict of interest.

References

1. Martin, J.W.; Scaglioni, B.; Norton, J.C.; Subramanian, V.; Arezzo, A.; Obstein, K.L.; Valdastri, P. Enabling the future of colonoscopy with intelligent and autonomous magnetic manipulation. *Nat. Mach. Intell.* **2020**, *2*, 595–606. [CrossRef] [PubMed]
2. Kumar, P.; Malik, S.; Toyserkani, E.; Khamesee, M.B. Development of an electromagnetic micromanipulator levitation system for metal additive manufacturing applications. *Micromachines* **2022**, *13*, 585. [CrossRef] [PubMed]
3. Cao, Q.; Fan, Q.; Chen, Q.; Liu, C.; Han, X.; Li, L. Recent advances in manipulation of micro-and nano-objects with magnetic fields at small scales. *Mater. Horizons* **2020**, *7*, 638–666. [CrossRef]
4. Yang, Z.; Zhang, L. Magnetic actuation systems for miniature robots: A review. *Adv. Intell. Syst.* **2020**, *2*, 2000082. [CrossRef]
5. Chen, X.Z.; Jang, B.; Ahmed, D.; Hu, C.; De Marco, C.; Hoop, M.; Mushtaq, F.; Nelson, B.J.; Pané, S. Small-scale machines driven by external power sources. *Adv. Mater.* **2018**, *30*, 1705061. [CrossRef] [PubMed]
6. Sitti, M.; Ceylan, H.; Hu, W.; Giltinan, J.; Turan, M.; Yim, S.; Diller, E. Biomedical applications of untethered mobile milli/microrobots. *Proc. IEEE* **2015**, *103*, 205–224. [CrossRef] [PubMed]
7. Xu, T.; Yu, J.; Yan, X.; Choi, H.; Zhang, L. Magnetic actuation based motion control for microrobots: An overview. *Micromachines* **2015**, *6*, 1346–1364. [CrossRef]
8. Yang, L.; Zhang, L. Motion control in magnetic microrobotics: From individual and multiple robots to swarms. *Annu. Rev. Control Robot. Auton. Syst.* **2021**, *4*, 509–534. [CrossRef]
9. Abbott, J.J.; Diller, E.; Petruska, A.J. Magnetic methods in robotics. *Annu. Rev. Control Robot. Auton. Syst.* **2020**, *3*, 57–90. [CrossRef]
10. Komaee, A.; Shapiro, B. Steering a ferromagnetic particle by optimal magnetic feedback control. *IEEE Trans. Control Syst. Technol.* **2011**, *20*, 1011–1024. [CrossRef]

11. Komaee, A. Feedback control for transportation of magnetic fluids with minimal dispersion: A first step toward targeted magnetic drug delivery. *IEEE Trans. Control Syst. Technol.* **2016**, *25*, 129–144. [CrossRef]
12. Fang, W.Z.; Xiong, T.; Pak, O.S.; Zhu, L. Data-driven intelligent manipulation of particles in microfluidics. *Adv. Sci.* **2023**, *10*, 2205382. [CrossRef]
13. Chapin, S.C.; Germain, V.; Dufresne, E.R. Automated trapping, assembly, and sorting with holographic optical tweezers. *Opt. Express* **2006**, *14*, 13095–13100. [CrossRef] [PubMed]
14. Cohen, A.E. Control of nanoparticles with arbitrary two-dimensional force fields. *Phys. Rev. Lett.* **2005**, *94*, 118102. [CrossRef]
15. Zhou, Q.; Sariola, V.; Latifi, K.; Liimatainen, V. Controlling the motion of multiple objects on a Chladni plate. *Nat. Commun.* **2016**, *7*, 12764. [CrossRef] [PubMed]
16. Probst, R.; Lin, J.; Komaee, A.; Nacev, A.; Cummins, Z.; Shapiro, B. Planar steering of a single ferrofluid drop by optimal minimum power dynamic feedback control of four electromagnets at a distance. *J. Magn. Magn. Mater.* **2011**, *323*, 885–896. [CrossRef] [PubMed]
17. Ongaro, F.; Pane, S.; Scheggi, S.; Misra, S. Design of an electromagnetic setup for independent three-dimensional control of pairs of identical and nonidentical microrobots. *IEEE Trans. Robot.* **2018**, *35*, 174–183. [CrossRef]
18. Liu, Y.; Feng, Y.; An, M.; Sarwar, M.T.; Yang, H. Advances in Finite Element Analysis of External Field-Driven Micro/Nanorobots: A Review. *Adv. Intell. Syst.* **2023**, *5*, 2200466. [CrossRef]
19. Weerasooriya, S.; El-Sharkawi, M.A. Identification and control of a dc motor using back-propagation neural networks. *IEEE Trans. Energy Convers.* **1991**, *6*, 663–669. [CrossRef]
20. Kolo, B.A. *Neural Networks in Magnetic Guidance*; University of Virginia: Charlottesville, VA, USA, 1998.
21. Yu, R.; Charreyron, S.L.; Boehler, Q.; Weibel, C.; Chautems, C.; Poon, C.C.; Nelson, B.J. Modeling electromagnetic navigation systems for medical applications using random forests and artificial neural networks. In Proceedings of the 2020 IEEE International Conference on Robotics and Automation (ICRA), Paris, France, 31 May–31 August 2020; IEEE: Piscataway, NJ, USA, 2020; pp. 9251–9256.
22. Kazemzadeh Heris, P.; Khamesee, M.B. Design and fabrication of a magnetic actuator for torque and force control estimated by the ann/sa algorithm. *Micromachines* **2022**, *13*, 327. [CrossRef]
23. Tariverdi, A.; Venkiteswaran, V.K.; Richter, M.; Elle, O.J.; Tørresen, J.; Mathiassen, K.; Misra, S.; Martinsen, Ø.G. A recurrent neural-network-based real-time dynamic model for soft continuum manipulators. *Front. Robot. AI* **2021**, *8*, 631303. [CrossRef]
24. Liu, J.; Wu, X.; Huang, C.; Manamanchaiyaporn, L.; Shang, W.; Yan, X.; Xu, T. 3-D autonomous manipulation system of helical microswimmers with online compensation update. *IEEE Trans. Autom. Sci. Eng.* **2020**, *18*, 1380–1391. [CrossRef]
25. Behrens, M.R.; Ruder, W.C. Smart Magnetic Microrobots Learn to Swim with Deep Reinforcement Learning. *Adv. Intell. Syst.* **2022**, *4*, 2200023. [CrossRef]
26. Salehi, M.; Nejat Pishkenari, H.; Zohoor, H. Position control of a wheel-based miniature magnetic robot using neuro-fuzzy network. *Robotica* **2022**, *40*, 3895–3910. [CrossRef]
27. Turan, M.; Ornek, E.P.; Ibrahimli, N.; Giracoglu, C.; Almalioglu, Y.; Yanik, M.F.; Sitti, M. Unsupervised odometry and depth learning for endoscopic capsule robots. In Proceedings of the 2018 IEEE/RSJ International Conference on Intelligent Robots and Systems (IROS), Madrid, Spain, 1–5 October 2018; IEEE: Piscataway, NJ, USA, 2018; pp. 1801–1807.
28. Norton, J.C.; Slawinski, P.R.; Lay, H.S.; Martin, J.W.; Cox, B.F.; Cummins, G.; Desmulliez, M.P.; Clutton, R.E.; Obstein, K.L.; Cochran, S.; et al. Intelligent magnetic manipulation for gastrointestinal ultrasound. *Sci. Robot.* **2019**, *4*, eaav7725. [CrossRef] [PubMed]
29. Ahmad, O.F.; Mori, Y.; Misawa, M.; Kudo, S.E.; Anderson, J.T.; Bernal, J.; Berzin, T.M.; Bisschops, R.; Byrne, M.F.; Chen, P.J.; et al. Establishing key research questions for the implementation of artificial intelligence in colonoscopy: A modified Delphi method. *Endoscopy* **2021**, *53*, 893–901. [CrossRef] [PubMed]
30. Fan, Q.; Zhang, P.; Qu, J.; Huang, W.; Liu, X.; Xie, L. Dynamic magnetic field generation with high accuracy modeling applied to magnetic robots. *IEEE Trans. Magn.* **2021**, *57*, 1–10. [CrossRef]
31. Barnoy, Y.; Erin, O.; Raval, S.; Pryor, W.; Mair, L.O.; Weinberg, I.N.; Diaz-Mercado, Y.; Krieger, A.; Hager, G.D. Control of Magnetic Surgical Robots With Model-Based Simulators and Reinforcement Learning. *IEEE Trans. Med. Robot. Bionics* **2022**, *4*, 945–956. [CrossRef]
32. Liu, D. The Application of Machine Learning for Designing and Controlling Electromagnetic Fields. Ph.D. Thesis, University of Wisconsin–Madison, Madison, WI, USA, 2021.
33. Huang, Z.; Zhu, J.; Shao, J.; Wei, Z.; Tang, J. Recurrent neural network based high-precision position compensation control of magnetic levitation system. *Sci. Rep.* **2022**, *12*, 11435. [CrossRef]
34. Charreyron, S.L.; Boehler, Q.; Kim, B.; Weibel, C.; Chautems, C.; Nelson, B.J. Modeling electromagnetic navigation systems. *IEEE Trans. Robot.* **2021**, *37*, 1009–1021. [CrossRef]
35. Garrido-Jurado, S.; Muñoz-Salinas, R.; Madrid-Cuevas, F.J.; Marín-Jiménez, M.J. Automatic generation and detection of highly reliable fiducial markers under occlusion. *Pattern Recognit.* **2014**, *47*, 2280–2292. [CrossRef]
36. Specht, D.F. A general regression neural network. *IEEE Trans. Neural Netw.* **1991**, *2*, 568–576. [CrossRef]
37. Pedregosa, F.; Varoquaux, G.; Gramfort, A.; Michel, V.; Thirion, B.; Grisel, O.; Blondel, M.; Prettenhofer, P.; Weiss, R.; Dubourg, V.; et al. Scikit-learn: Machine Learning in Python. *J. Mach. Learn. Res.* **2011**, *12*, 2825–2830.
38. *Deep Learning Toolbox*, Version: 9.4 (R2022b); The MathWorks Inc.: Natick, MA, USA, 2022.

39. Moré, J.J. The Levenberg-Marquardt algorithm: Implementation and theory. In *Numerical Analysis, Proceedings of the Biennial Conference, Dundee, UK, June 28–July 1 1977*; Springer: Berlin/Heidelberg, Germany, 2006; pp. 105–116.
40. Guennebaud, G.; Jacob, B.; Avery, P.; Bachrach, A.; Barthelemy, S.; Becker, C.; Benjamin, D.; Berger, C.; Berres, A.; Blanco, J.L.; Borgerding, M.; Bossart, R.; et al. Eigen v3. 2010. Available online: http://eigen.tuxfamily.org (accessed on 15 January 2019).
41. Galan, E.A.; Zhao, H.; Wang, X.; Dai, Q.; Huck, W.T.; Ma, S. Intelligent microfluidics: The convergence of machine learning and microfluidics in materials science and biomedicine. *Matter* **2020**, *3*, 1893–1922. [CrossRef]
42. Shapiro, B.; Kulkarni, S.; Nacev, A.; Muro, S.; Stepanov, P.Y.; Weinberg, I.N. Open challenges in magnetic drug targeting. *Wiley Interdiscip. Rev. Nanomed. Nanobiotechnol.* **2015**, *7*, 446–457. [CrossRef] [PubMed]
43. Vitol, E.A.; Novosad, V.; Rozhkova, E.A. Microfabricated magnetic structures for future medicine: From sensors to cell actuators. *Nanomedicine* **2012**, *7*, 1611–1624. [CrossRef] [PubMed]
44. Shao, Y.; Fahmy, A.; Li, M.; Li, C.; Zhao, W.; Sienz, J. Study on magnetic control systems of micro-robots. *Front. Neurosci.* **2021**, *15*, 736730. [CrossRef] [PubMed]
45. Hong, A.; Petruska, A.J.; Zemmar, A.; Nelson, B.J. Magnetic control of a flexible needle in neurosurgery. *IEEE Trans. Biomed. Eng.* **2020**, *68*, 616–627. [CrossRef]
46. Kinross, J.M.; Mason, S.E.; Mylonas, G.; Darzi, A. Next-generation robotics in gastrointestinal surgery. *Nat. Rev. Gastroenterol. Hepatol.* **2020**, *17*, 430–440. [CrossRef] [PubMed]
47. Connor, M.J.; Dasgupta, P.; Ahmed, H.U.; Raza, A. Autonomous surgery in the era of robotic urology: Friend or foe of the future surgeon? *Nat. Rev. Urol.* **2020**, *17*, 643–649. [CrossRef] [PubMed]
48. Brablc, M.; Žegklitz, J.; Grepl, R.; Babuška, R. Control of Magnetic Manipulator Using Reinforcement Learning Based on Incrementally Adapted Local Linear Models. *Complexity* **2021**, *2021*, 6617309. [CrossRef]
49. Abbasi, S.A.; Ahmed, A.; Noh, S.; Gharamaleki, N.L.; Kim, S.; Chowdhury, A.M.B.; Kim, J.y.; Pané, S.; Nelson, B.J.; Choi, H. Autonomous 3D positional control of a magnetic microrobot using reinforcement learning. *Nat. Mach. Intell.* **2024**, *6*, 92–105. [CrossRef]
50. Cai, M.; Wang, Q.; Qi, Z.; Jin, D.; Wu, X.; Xu, T.; Zhang, L. Deep Reinforcement Learning Framework-Based Flow Rate Rejection Control of Soft Magnetic Miniature Robots. *IEEE Trans. Cybern.* **2022**, *53*, 7699–7711. [CrossRef] [PubMed]

robotics

MDPI

Article

Learning Advanced Locomotion for Quadrupedal Robots: A Distributed Multi-Agent Reinforcement Learning Framework with Riemannian Motion Policies

Yuliu Wang [1,2,*], Ryusuke Sagawa [1,2] and Yusuke Yoshiyasu [2]

[1] Intelligent and Mechanical Interaction System, University of Tsukuba, Tsukuba 305-8577, Ibaraki, Japan; ryusuke.sagawa@aist.go.jp
[2] Computer Vision Research Team, Artificial Intelligence Research Center, The National Institute of Advanced Industrial Science and Technology, 1-1-1 Umezono, Tsukuba 305-8560, Ibaraki, Japan; yusuke-yoshiyasu@aist.go.jp
* Correspondence: s2130176@u.tsukuba.ac.jp

Abstract: Recent advancements in quadrupedal robotics have explored the motor potential of these machines beyond simple walking, enabling highly dynamic skills such as jumping, backflips, and even bipedal locomotion. While reinforcement learning has demonstrated excellent performance in this domain, it often relies on complex reward function tuning and prolonged training times, and the interpretability is not satisfactory. Riemannian motion policies, a reactive control method, excel in handling highly dynamic systems but are generally limited to fully actuated systems, making their application to underactuated quadrupedal robots challenging. To address these limitations, we propose a novel framework that treats each leg of a quadrupedal robot as an intelligent agent and employs multi-agent reinforcement learning to coordinate the motion of all four legs. This decomposition satisfies the conditions for utilizing Riemannian motion policies and eliminates the need for complex reward functions, simplifying the learning process for high-level motion modalities. Our simulation experiments demonstrate that the proposed method enables quadrupedal robots to learn stable locomotion using three, two, or even a single leg, offering advantages in training speed, success rate, and stability compared to traditional approaches, and better interpretability. This research explores the possibility of developing more efficient and adaptable control policies for quadrupedal robots.

Keywords: multi-agent reinforcement learning; Riemannian motion policies; motion control; robot learning; quadrupedal robotics

1. Introduction

1.1. Background

Quadrupedal robots have garnered significant attention in recent years due to their potential for navigating complex terrains and performing versatile tasks. Compared to wheeled or humanoid robots, quadrupedal robots combine more stability with flexible locomotion. Quadrupedal motion is a widespread structure in nature, and quadrupedal robots have a lot of potential to improve their locomotion. Recent advances in quadrupedal robotics have focused on improving locomotion control, enabling these robots to perform more dynamic and agile maneuvers. Traditional approaches to quadrupedal locomotion control often relied on hand-crafted algorithms and control strategies, such as the model predictive control (MPC) [1] or the zero-moment point (ZMP) methods [2]. These methods have been successful in achieving stable walking on flat surfaces and have been implemented on various quadrupedal platforms, such as the MIT Cheetah and the ANYmal robot. However, these approaches struggle to adapt to uneven terrains and lack the flexibility to perform more dynamic maneuvers, such as jumping or running. To address the limitations of traditional approaches, machine learning techniques have been increasingly applied to quadrupedal robot control. Supervised learning methods, such as neural networks, have been used to learn locomotion patterns from expert demonstrations. These methods have shown promising results in generating stable and efficient gaits for quadrupedal robots.

Reinforcement learning (RL) has emerged as a promising approach for enabling quadrupedal robots to learn adaptive and versatile locomotion skills [3]. By allowing the robot to learn through trial-and-error interactions with its environment, RL can discover optimal control policies without relying on explicit programming. Recent advancements in deep reinforcement learning (DRL) have further enhanced the capabilities of quadrupedal robots, enabling them to learn complex behaviors such as jumping, backflips, and even bipedal walking [4–6]. Enabling quadrupedal robots to walk on three or two legs has significant application value, as it allows the free legs of quadrupedal robots to perform manipulation tasks, combining the advantages of both quadrupedal and humanoid robots. This capability extends the functionality of quadrupedal robots, making them more versatile and adaptable to various scenarios, such as search and rescue operations, industrial inspection, and space exploration.

Despite the successes of DRL in quadrupedal locomotion, several challenges remain. One major issue is the need for carefully designed reward functions that guide the learning process toward desired behaviors. Crafting effective reward functions often requires extensive domain knowledge and trial-and-error tuning, which can be time consuming and limit the scalability of these methods. As the robot learns more complex skills, the design of the reward function becomes increasingly complex and task-specific, leading to less generalizable policies. This lack of generalizability hinders the robot's ability to adapt to new tasks or environments without extensive retraining. Additionally, the learned policies often lack interpretability, making it difficult to understand and trust the robot's decision-making process.

Riemannian motion policies (RMPs) have emerged as a promising approach for generating reactive and adaptive motions in highly dynamic environments. RMPs leverage the geometric structure of the robot's configuration space to produce smooth and efficient motions that can quickly respond to changes in the environment [7]. The key principle behind RMPs is to define a Riemannian metric that encodes the desired behavior of the robot, such as obstacle avoidance or goal reaching, and then, use this metric to generate motion commands. One of the main advantages of RMPs is their ability to output reactive actions, which are crucial for robots operating in dynamic and uncertain environments. By continuously updating the robot's motion based on the current state and the learned metric, RMPs can generate appropriate actions in real time without the need for extensive planning or optimization. This reactivity makes RMPs well suited for handling disturbances and adapting to changing conditions. However, the application of RMPs has primarily been limited to fully actuated systems [8], such as robotic arms [9]. In a fully actuated system, the number of actuators (motors or joints) is equal to the number of degrees of freedom, allowing the robot to independently control each degree of freedom. This property simplifies the application of RMPs, as the learned metric can directly map to the robot's actuation commands.

On the other hand, many robotic systems, including quadrupedal robots, are underactuated, meaning they have fewer actuators than degrees of freedom. In these systems, the direct application of RMPs becomes more challenging, as the robot cannot independently control all its degrees of freedom. To address this issue, some studies have explored the use of kinematic decomposition to apply RMPs to underactuated systems, such as wheeled inverted pendulum robots [10]. By decomposing the robot's motion into controllable and uncontrollable subspaces, these approaches can apply RMPs to the controllable subspace while handling the uncontrollable dynamics separately. Recent research has also investigated the application of RMPs to bipedal robots, which are inherently underactuated due to their floating base [11]. These studies have shown promising results in generating stable and adaptive walking motions by combining RMPs with other control techniques, such as whole-body control or model predictive control. However, the application of RMPs to quadrupedal robots, which exhibit more complex dynamics and require coordination between multiple legs, remains an open challenge.

1.2. Contributions

We propose a novel method for a quadrupedal robot to learn advanced motions, such as three-legged, two-legged, and one-legged locomotion (Figure 1). Our hierarchical architecture integrates multi-agent reinforcement learning (MARL) with Riemannian motion policies (RMPs) and a multi-head structure (Figure 2). The main contributions of our work are:

- Extending the applicability of RMPs to underactuated systems, such as quadrupedal robots, by leveraging the MARL framework for dynamic space decomposition.
- Learning diverse locomotion modes in a unified framework, such as three-legged, two-legged, and one-legged walking, within a single architecture, enabling the robot to maintain balance and stability in different motion policies by performing reactive center-of-mass tracking.
- Introduction of a multi-head structure agent to handle both discrete and continuous actions simultaneously.
- Demonstration of the method's effectiveness through simulation experiments, showcasing advantages in learning efficiency, stability, and success rate, particularly in challenging tasks.

Figure 1. Our proposed method enables quadrupedal robots to quickly learn advanced skills such as bipedal walking.

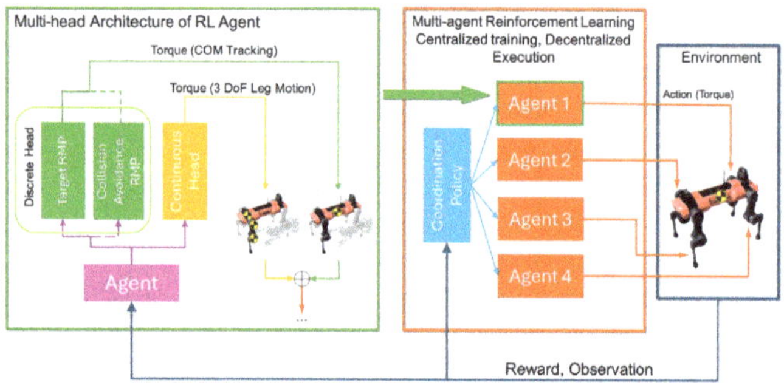

Figure 2. Overview: The system consists of two main components: the multi-head architecture of the RL agent with RMPs, and the multi-agent reinforcement learning framework for centralized training and decentralized execution. The combination of the multi-head architecture and multi-agent reinforcement learning allows the system to learn complex and adaptive locomotion behaviors for the quadrupedal robot. The multi-head architecture enables the generation of appropriate torques for each leg, considering factors such as motion mode, self-collision avoidance, and center-of-mass tracking. The multi-agent reinforcement learning framework facilitates the coordination and synchronization of the four legs to achieve stable and efficient locomotion.

1.3. Layout

This paper is organized as follows. In Section 2, the fundamentals of the techniques used in our proposed framework are presented. This section includes the theory of Riemannian motion policies and the basic principles of reinforcement learning to control quadrupedal robots (Sections 2.1–2.3), and the working principle of distributed multi-agent reinforcement learning. In Section 3, how to build our proposed framework is presented, including the design of the RMPs, the design and implementation of agents with multi-head structure, and training using the distributed multi-agent RL framework (Sections 3.1–3.4). In Section 4, experimental results training, simulation, and quantitative analysis are presented (Sections 4.1–4.3). Finally, we summarize the methodology and results in Section 5.1 and discuss the strengths and weaknesses and future work in Section 5.2.

2. Related Work

2.1. RMPs and RMPflow

Riemannian motion policies (RMPs), introduced by Ratliff and colleagues in [8], provide a novel perspective on motion planning and control by viewing task space as a manifold and describing motion using geodesics, which are the shortest paths on this manifold. In this framework, motion policies are considered generators of geodesics, and their behavior can be controlled by altering the Riemannian metric of the manifold. By choosing appropriate metrics, geodesics can form curved trajectories, exemplifying the application of RMPs.

Formally, RMPs are specified for a manifold $\mathcal{M}$ with coordinate map x and defined acceleration $\ddot{x} = a(x, \dot{x})$, where a is the motion policy. An RMP is represented as $(a, M)^{\mathcal{M}}$, which combines the motion policy a with an importance weight M, a symmetric positive-definite matrix function that often represents the Riemannian metric of the manifold. The expression $(a, M)^{\mathcal{M}}$ is known as the canonical form of an RMP, while $[f, M]^{\mathcal{M}}$ is referred to as the natural form, with $f := Ma$ being the control force. One of the key advantages of RMPs is that motion policies can be transformed between manifolds using techniques from differential geometry, treating them as geometric objects. This allows for the seamless integration of multiple motion policies defined on different task manifolds. The configuration space $\mathcal{C}$ of a robot is an n-dimensional manifold with a global chart $(\mathcal{C}, q)$, and task manifold motion policies can be mapped onto $\mathcal{C}$ using techniques like the pullback operation. By leveraging the geometric structure of the task space and the configuration space, RMPs offer a powerful tool for motion planning and control in various applications, such as robotic manipulation and locomotion.

RMPflow [12] is a computational framework that embodies this concept, facilitating the consistent integration of task-space RMPs into a single motion policy on $\mathcal{C}$. It employs a tree-structured computational graph called an RMP-tree. Within an RMP-tree, each node represents an RMP and its corresponding state on a manifold, while the edges signify the operations that transform RMPs and states between nodes. The root node corresponds to the configuration space, and the leaf nodes correspond to the task spaces. The set of edge operations, collectively termed RMP-algebra, consists of three transformations: pushforward, pullback, and resolve.

2.2. Deep Reinforcement Learning for Quadrupedal Robots

Deep reinforcement learning (DRL) has emerged as a powerful approach for enabling quadruped robots to learn complex locomotion skills [13]. In DRL, an agent learns to make optimal decisions by interacting with its environment and receiving rewards or penalties based on its actions. The goal is to maximize the cumulative reward over time, which encourages the agent to discover effective control policies.

To apply DRL to quadruped robot control, the problem is typically modeled as a Markov decision process (MDP). An MDP is defined by a tuple $(\mathcal{S}, \mathcal{A}, \mathcal{P}, \mathcal{R}, \gamma)$, where $\mathcal{S}$ represents the state space, $\mathcal{A}$ denotes the action space, $\mathcal{P}$ describes the transition probability between states, $\mathcal{R}$ is the reward function, and $\gamma \in [0, 1]$ is the discount factor that determines the importance of future rewards. In the context of quadruped robot control, the state space $\mathcal{S}$ typically includes information such as the robot's joint angles, velocities, and body orientation. The action space $\mathcal{A}$ corresponds to the motor commands sent to the robot's actuators, which can be continuous (e.g., joint torques) or discrete (e.g., predefined gait patterns). The transition probability $\mathcal{P}$ represents the dynamics of the robot and its interaction with the environment, which can be either explicitly modeled or learned through experience.

The reward function $\mathcal{R}$ plays a crucial role in guiding the learning process. It assigns a scalar value to each state–action pair, indicating the desirability of the action taken in that state. In quadruped locomotion tasks, the reward function often includes terms that encourage forward progress, stability, and energy efficiency, while penalizing falls and undesired behaviors [14].

A typical DRL training process for quadruped robot locomotion involves the following steps: The agent observes the current state s_t of the robot and its environment. Based on the observed state, the agent selects an action a_t according to its policy $\pi(a_t|s_t)$. The robot executes the chosen action a_t and transitions to a new state s_{t+1} according to the transition probability $\mathcal{P}(s_{t+1}|s_t, a_t)$. The agent receives a reward $r_t = \mathcal{R}(s_t, a_t)$ based on the action taken and the resulting state. The agent updates its policy π using a DRL algorithm, such as Q-learning [15], policy gradients [16], or actor–critic methods [17], to maximize the expected cumulative reward. The process repeats from step 1 until the agent converges to a satisfactory policy or a predefined termination criterion is met.

By iteratively exploring the state–action space and updating its policy, the DRL agent learns to make intelligent decisions that enable the quadruped robot to achieve stable and efficient locomotion. However, as the complexity of the task increases, the design of the reward function becomes more and more complex. For example, detailed objectives for the positions that need to be achieved by each joint of the robot.

2.3. Distributed Multi-Agent Reinforcement Learning Framework

Multi-agent reinforcement learning (MARL) is an extension of single-agent reinforcement learning (SARL) that deals with the challenges of learning in environments where multiple agents interact [18]. In MARL, each agent aims to learn an optimal policy that maximizes its individual or shared reward while considering the actions of other agents. The presence of multiple agents introduces additional complexities, such as coordination, communication, and potential conflicts among agents.

The key difference between MARL and SARL lies in the environment and the interactions among agents. In SARL, a single agent interacts with the environment and aims to maximize its cumulative reward. The agent's actions directly influence the environment, and the environment's state transitions depend solely on the agent's actions. In contrast, MARL involves multiple agents acting simultaneously in the same environment. Each agent's actions not only affect the environment but also influence the actions and rewards of other agents [19]. This interdependence among agents necessitates consideration of joint actions, joint observations, and potentially conflicting goals.

Diverging from the singular entity focus of traditional reinforcement learning and the decision-making framework of Markov decision processes, the domain of multi-agent reinforcement learning is encapsulated by the dynamics of a Markov game. This scenario unfolds with several agents independently and concurrently making decisions and taking actions, steered by the prevailing environmental state. The cumulative action choices determine the trajectory of environmental progression and its evolving state. The structure of this Markov game is encapsulated by the tuple $(S, A_1 ... A_i, T, R_1 ... R_i, \gamma)$, where S is the inventory of possible states, A_i and R_i are the respective sets of possible actions and rewards for agent i, and T embodies the state transition probabilities alongside the discount factor γ. The expected return for agent i, when employing policy π^i, is quantified as

$$V^{\pi^i, \pi^{-i^i}} := E\left[\sum t \geq 0 \gamma^t R^i(s_t, a_t, s_{t+1}) | a_t^i \sim \pi^i(\cdot|s_t), s_0 = s\right] \quad (1)$$

In multi-agent reinforcement learning pursuits, the quest is to achieve a Nash equilibrium among all agents involved. This conveys that in a collective scenario where agents cooperate, no single agent can improve its aggregate rewards by unilaterally altering its established policy. It is imperative to note that the Nash equilibrium is not synonymous with the absolute optimum; rather, it epitomizes the most expected stable condition and represents a convergence target during the learning phase.

Our previous work [20] has demonstrated that MARL can be effectively applied to train individual robots by treating the robot's individual driving parts as separate agents. This approach offers several advantages. First, it compresses the action space, as each agent is responsible for controlling a specific part of the robot, rather than the entire robot. Second, it speeds up the training process, as the agents can learn their policies in parallel. Third, it provides each part of the robot with some degree of autonomy, allowing for more flexible and reactive behavior.

3. Methods

3.1. Overview

To realize our approach, we design a hierarchical architecture that integrates multi-agent reinforcement learning with RMPs. The top layer consists of collaborative policies that coordinate the movement of the four legs, while in the middle layer, we build an agent using a multi-head structure, which can output both discrete and continuous actions. In the bottom layer, multiple RMPs are deployed within the discrete head, which executes center-of-mass (COM) tracking actions based on the selected RMPs, the continuous head then outputs leg torque to ensure safety and stability during locomotion. In the following, we will detail how each layer is constructed.

3.2. RMPs for Leg Locomotion Control

In our analytical treatment of the quadruped robot's dynamics, we initially partition the system into four discrete dynamical sectors. Each sector encompasses an individual leg in conjunction with the robot's central trunk. Within each of these independent sectors, we identify three articulations corresponding to three degrees of freedom, thereby constituting a fully actuated subsystem, shown in Figure 3. This segmentation facilitates the direct application of RMPs to each leg–trunk pair.

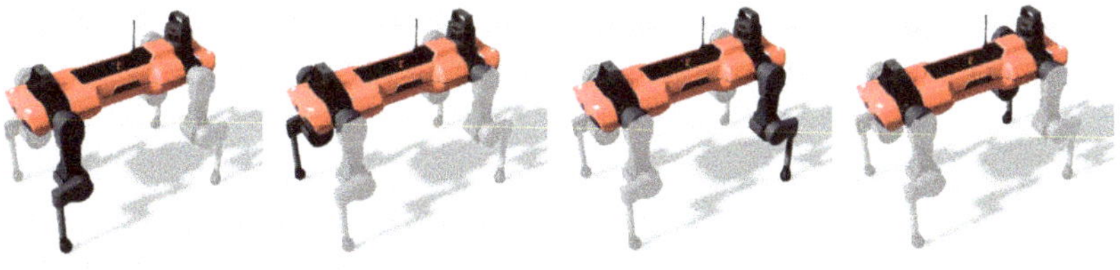

(**a**) Agent 1 (**b**) Agent 2 (**c**)Agent 3 (**d**) Agent 4

Figure 3. The area controlled by each agent includes a leg and trunk section.

It is critical to acknowledge that the trunk segment exhibits a state of superposition across the four delineated spaces. This characteristic is harmonious with the underlying architecture of our proposed model, wherein the collective operation of the quadruped's limbs is synchronized to fortify the trunk's stability. This synergy is a cornerstone of our approach, ensuring that the locomotive integrity of the robot is preserved through the coordinated mechanics of its constituent parts. The dynamics of legged robotic systems can be encapsulated by a set of fundamental equations that describe their motion:

$$\mathbf{A}\ddot{\mathbf{q}} + \mathbf{b} + \mathbf{g} = \mathbf{S}_a^\top \boldsymbol{\tau} + \boldsymbol{J}_c^\top \mathbf{f}_r \tag{2}$$

In this equation, A represents the generalized mass matrix, $\mathbf{b}$ stands for the Coriolis force vector, and $\mathbf{g}$ is the vector of gravitational forces. The matrix S_a is used to select the actuated joints. The vectors $\boldsymbol{\tau}$ and $\mathbf{f}_r$ denote the torques at the joints and the augmented reaction forces, respectively, while J_c is the contact Jacobian matrix that relates the forces at the points of contact to the joint torques. The term $\ddot{\mathbf{q}} \in \mathbb{R}^{6+n}$ signifies the acceleration within the configuration space, with n being the count of actuated degrees of freedom in the robot's structure. In this section, we will use two RMPs for control: the target RMP and the collision avoidance RMP.

3.2.1. Target RMP for Center-of-Mass Tracking

Center-of-mass tracking is an effective policy for keeping quadrupedal robots stable [21]. The function of the target RMP is to maintain the COM of the robot's trunk at the desired position by outputting the leg's locomotion. This function is similar to the target RMP in a robotic arm. However, there is a key difference: the robot's foot contact with the ground is not fixed. This variability can be considered a kind of disturbing noise. This issue is addressed later on in reinforcement learning. In addition, it is almost impossible for one leg to maintain the center-of-mass stabilization on its own, but the four-legged agents can work together to solve this problem, and here, we only need to formulate a local RMP. The target RMP's behavior is governed by the following equation:

$$\ddot{x} = k_p(x_0 - x)/(|x_0 - x| + \epsilon) - k_d|\dot{x}|\dot{x} \tag{3}$$

The term k_p represents the position gain, which plays a crucial role in the control of motion policies. The damping RMP, on the other hand, is given by $-k_d|\dot{x}|\dot{x}$, where k_d is the damping gain. This damping term is essential because undamped Riemannian motion policies can lead to persistent oscillations when tracking a dynamic target. By incorporating damped RMPs, these oscillations can be significantly reduced, resulting in smoother and more stable motion. The parameter ϵ sets the transition scale from constant acceleration to linear behavior around the target. This parameter allows for a smooth transition between different motion regimes, ensuring that the robot's movement remains continuous and well behaved.

Each motion policy in the RMP framework is associated with a specific Riemannian metric, denoted by M. The Riemannian metric plays a crucial role in determining the behavior of the motion policy, as it defines the notion of distance and curvature on the manifold. For a comprehensive understanding of how the Riemannian metric is derived, the reader is referred to [8], which provides a detailed explanation of the underlying mathematics. The Riemannian metric for the target RMP is constructed as follows:

$$M = [\beta(x)b + (1 - \beta(x))][\alpha(x)M_{\text{near}} + (1 - \alpha(x))M_{\text{far}}] \tag{4}$$

The metric M, also known as the inertia matrix, lies within the range defined by the rank-deficient metric S and the identity matrix I. The metric S has a significant impact in areas far from the target, while the identity matrix I becomes dominant near the target, promoting faster convergence. The blending of these matrices is controlled by a radial basis function, specifically a Gaussian function, which smoothly transitions from a constant minimum value far from the target to a value close to 1 near the target.

In the above equations, the following definitions apply:

$$\begin{aligned}
&\alpha(x) = (1 - \alpha_{\min})\exp\left(\frac{-||x_0 - x||^2}{2\sigma_a^2}\right) + \alpha_{\min} \\
&\beta(x) = \exp\left(-\frac{||x_0 - x||^2}{2\sigma_b^2}\right) \\
&M_{\text{near}} = \mu_{\text{near}} I \\
&M_{\text{far}} = \mu_{\text{far}} S = \frac{\mu_{\text{far}}}{||x_0 - x||^2}(x_0 - x)(x_0 - x)^T
\end{aligned} \tag{5}$$

The parameters σ_a and σ_b are length scales, measured in meters, for the Gaussian functions used in the construction of the Riemannian metric for the target RMP. Specifically, σ_a is the length scale for the Gaussian function that blends the matrices S and I, which are used to define the metric's behavior. On the other hand, σ_b is the length scale for the Gaussian function that enhances boosting near the target, allowing for more aggressive control when the robot is close to its goal. The parameter $\alpha_{\min}$ sets the minimum influence of M_{near} on the overall Riemannian metric M. This ensures that the metric maintains a certain level of responsiveness even when the robot is far from the target. The priority weights μ_{near} and μ_{far} determine the relative influence of M_{near} and M_{far} on the overall metric M, respectively. These weights allow for a balance between the near and far target behaviors, enabling the robot to smoothly transition between different control regimes. Finally, the scaling factor b adjusts the intensity of the strengthening effect near the target. A higher value of b results in a more pronounced boosting effect, allowing for faster convergence to the target when the robot is in close proximity.

3.2.2. Collision Avoidance RMP for Leg Locomotion

The RMP framework includes a collision avoidance RMP designed to prevent collisions between the robot's legs. To ensure computational efficiency, the geometric properties of the robot's links are approximated using capsules, reducing the collision avoidance problem to avoiding collisions between witness points on these capsules. Only one collision RMP is required for each pair of link capsules, and the position of the witness points is determined using the algorithm proposed in the RMP framework. These collision avoidance RMPs are defined in a one-dimensional Euclidean distance space and are characterized by a repulsive acceleration policy and a corresponding metric. The repulsive acceleration policy ensures that the witness points on the capsules are pushed away from each other when they come too close, while the corresponding metric determines the strength and shape of the repulsive force, allowing for the desired balance between safety and efficiency.

$$\begin{aligned} \ddot{x}(x,\dot{x}) &= k_p \exp(-x/l_p) - k_d \frac{\sigma(\dot{x})\dot{x}}{x/l_d + \epsilon_d} \\ m(x,\dot{x}) &= \sigma(\dot{x}) g(x) \frac{\mu}{x/l_m + \epsilon_m} \end{aligned} \tag{6}$$

where

$$\begin{aligned} \sigma(\dot{x}) &= 1 - \frac{1}{1+\exp(-\dot{x}/v_d)} \\ g(x) &= \begin{cases} x^2/r^2 - 2x/r + 1, & x \le r \\ 0, & x > r \end{cases} \end{aligned} \tag{7}$$

The collision avoidance RMP is defined in terms of the Euclidean distance x between witness points on the capsules and its rate of change $\dot{x}$. The repulsion gain $k_p\left[\mathrm{m/s^2}\right]$ and damping gain $k_d\left[s^{-1}\right]$ determine the strength of the repulsive force and resistance to movement, respectively. The parameter μ sets the relative priority of the collision RMP, which is adjusted end-to-end by reinforcement learning (RL) in subsequent layers. The threshold distance $r[m]$ determines the activation of the collision RMP, while scaling parameters $l_p[m]$, $l_m[m]$, and $v_d[m/s]$ and offset parameters ϵ_m and ϵ_d enable fine-tuning of the collision avoidance behavior based on the specific requirements of the task and the robot's environment.

3.3. Multi-Head Architecture of RL Agent

As discussed in the previous sections, we have decomposed the quadrupedal robot into four dynamic sections, each controlled by an agent with a multi-head structure [22]. This approach allows for more efficient and effective training of the robot, as each agent is responsible for controlling a specific part of the robot and can learn its policy.

The multi-head structure of the reinforcement learning agents enables them to output a mixture of actions, both discrete and continuous. The reason we introduce a multi-head structure is that we are not planning for gait and only need to satisfy the final reward maximization, so the movement of the legs is completely controlled end-to-end by the policy network. In this study, we employ two heads, a discrete head and a continuous head, and the two heads share an environment feature extractor. The discrete head is responsible for deciding whether to use the target RMP or the collision avoidance RMP. This decision is made within a short time frame of less than 2 ms, ensuring reactive action output. The moments generated by the selected RMP are then tracked against the COM position of the robot's trunk, as the task requires a quick response to deviations.

Multi-head reinforcement learning (RL) and hierarchical RL are similar in structure, but they differ significantly in how they operate. Multi-head RL simultaneously generates multiple actions, whereas hierarchical RL chooses just one action from several options before executing it. This is a Markov decision process (MDP) characterized by a set of possible states $s \in \mathcal{S}$, a repertoire of actions $a \in \mathcal{A}$, a probability distribution for state transitions $s_{t+1} \sim P(s_t, a_t)$, and a function that assigns rewards $r(s, a)$. We refer to $\pi = \pi(a \mid s), s \in \mathcal{S}, a \in \mathcal{A}$ as a stochastic policy. The structure of the multi-head reinforcement learning agent can be described as follows:

$$J(\pi(a|s)) = \begin{cases} J(\pi_d(a_d|s)) \triangleq \underset{\tau\sim\pi_d}{\mathbb{E}}\left[\sum_{t=0}^{\infty}\gamma^t r(s_t, a_d = [0,1])\right] & \text{for discrete actions} \\ \\ J(\pi_c(a_c|s)) \triangleq \underset{\tau\sim\pi_c}{\mathbb{E}}\left[\sum_{t=0}^{\infty}\gamma^t r(s_t, a_c)\right] & \text{for continuous actions} \end{cases} \tag{8}$$

where $\pi(a|s)$ represents the overall policy of the agent, $\pi_d(a_d|s)$ denotes the policy for the discrete head, and $\pi_c(a_c|s)$ represents the policy for the continuous head. The discrete action a_d is sampled from a categorical distribution, while the continuous action a_c is sampled from a continuous distribution, such as a Gaussian distribution. The discrete head of the agent selects between the target RMP and the collision avoidance RMP based on the current state s. This selection is made by sampling from the categorical distribution output by the discrete policy $\pi_d(a_d|s)$. The chosen RMP then generates the appropriate moments to be tracked against the COM position of the trunk.

The discount factor γ, which lies between 0 and 1 (not inclusive), reduces the value of future rewards, and a trajectory $\tau = (s_0, a_0, s_1, \ldots)$ represents a sequence of states and actions obtained from the system. The strategy of policy gradient techniques is to refine the policy in a direct manner. Typically, the policy π is realized using a function that is defined by parameters θ. Moving forward, we will use θ to refer both to the specific parameters and the policy itself. In the domain of DRL, this policy is often encoded within a neural network known as the policy network, and we assume that it is smoothly differentiable with respect to the parameters θ.

On the other hand, the continuous head of the agent outputs moments for each joint of the leg with 3 degrees of freedom (DOF). These moments are used to control the leg motion and are sampled from the continuous distribution output by the continuous policy $\pi_c(a_c|s)$. The continuous actions are then applied to the corresponding joints of the leg, enabling smooth and coordinated locomotion.

The multi-head structure of the reinforcement learning agent allows for the simultaneous learning of discrete and continuous actions, which is particularly useful in tasks that require both high-level decision making and fine-grained control. By decomposing the control of the quadrupedal robot into separate agents with multi-head structures, we can effectively train the robot to perform complex tasks that involve both reactive decision making and precise leg motion control.

3.4. Updating Policies Using Distributed Multi-Agent RL Framework

Having constructed the agents using a multi-head structure, we proceed to train the four agents using distributed multi-agent reinforcement learning. We employ a centralized training, decentralized execution (CTDE) paradigm, specifically, the Multi-Agent Proximal Policy Optimization (MAPPO) method [23]. CTDE [24] is a popular approach in multi-agent reinforcement learning, where the training process is centralized, allowing agents to access additional information and learn coordinated policies, while the execution is decentralized, enabling each agent to make decisions based on its local observations. This paradigm strikes a balance between the benefits of centralized learning and the practicality of decentralized execution. MAPPO is an extension of the Proximal Policy Optimization (PPO) algorithm [25] to the multi-agent setting. In MAPPO, each agent maintains its policy π_i and value function V_i, while a central critic V_{tot} is used to estimate the total expected return of all agents. The central critic takes the joint state s and joint action $\mathbf{a} = (a_1, \ldots, a_n)$ as input, where n is the number of agents. During training, MAPPO updates the policies of all agents simultaneously using the following objective function:

$$L(\theta_i) = \\ \mathbb{E}s, \mathbf{a} \sim \pi\theta_{old}\left[\min\left(\frac{\pi_{\theta_i}(a_i|s)}{\pi_{\theta_{old,i}}(a_i|s)}A^{MAPPO}(s,\mathbf{a}), \text{clip}\left(\frac{\pi_{\theta_i}(a_i|s)}{\pi_{\theta_{old,i}}(a_i|s)}, 1-\epsilon, 1+\epsilon\right)A^{MAPPO}(s,\mathbf{a})\right)\right] \tag{9}$$

where θ_i and $\theta_{old,i}$ are the parameters of the current and previous policies of agent i, respectively, and $A^{MAPPO}(s, \mathbf{a})$ is the multi-agent advantage function, defined as

$$A^{MAPPO}(s,\mathbf{a}) = r + \gamma V_{tot}(s', \mathbf{a}') - V_{tot}(s, \mathbf{a}) \tag{10}$$

where r is the immediate reward, γ is the discount factor, and s' and $\mathbf{a}'$ are the next state and joint action, respectively.

In our task, the other legs are part of the environment for the leg being controlled, which provides a stable training environment for each individual agent. The central coordination policy helps to adjust the weights of each agent's actions in different states, all agents can be in a state of dynamic equilibrium through collaboration. Specifically, for an agent, the action outputs of all other agents become environmental perturbations for this agent, who maintains the output action in the center-of-mass position. Similarly the action output of agent will also perturb the environment of all other agents, which in turn will allow all agents to make adjustments. The collaborative mechanism generated by the centralized training and reward function will help the multiple agents to enter a state of dynamic equilibrium. While a single agent cannot maintain a stable COM on its own, the coordination of all four agents can achieve this goal. Therefore, we do not need to pre-train individual agents and can proceed directly to full end-to-end training to obtain the desired result.

By utilizing the CTDE paradigm and the MAPPO algorithm, we can effectively train the multi-head agents to learn coordinated policies that enable the quadrupedal robot to maintain a stable COM and perform the desired tasks. The centralized training allows the agents to learn cooperative behaviors, while the decentralized execution ensures that each agent can make decisions based on its local observations during deployment.

3.5. Rewards

The reward function is a crucial component of reinforcement learning, as it guides the policy update process by providing feedback on the desirability of the agent's actions. The complexity of the reward function often increases with the complexity of the task. For example, some previous studies have designed detailed reward terms to approximate the robot's actions to teacher samples through adversarial and imitation learning [3,6]. For a standing task, the reward function needs to be designed as follows:

$$R = r_\alpha + r_{height} + r_{\text{feet}} + r_{wheels} + r_{\text{shoulder}} + r_{\text{stand pose}} \tag{11}$$

where

$$\begin{aligned} r_\alpha &= \frac{\pi/2 - \alpha}{\pi/2}, r_{height} = p_z^{\text{robot}} \\ r_{\text{feet}} &= f, r_{wheels} = \sum \dot{q}^2_{frontwheels} \times (1 - f) \\ r_{\text{shoulder}} &= \|q_{\text{shoulder}}\|^2, r_{\text{stand pose}} = \exp\left(-0.1 \times \|q_{hl} - q_{0,hl}\|^2\right) \end{aligned} \tag{12}$$

Here, the $q^{\text{robot}} \in \mathbb{H}$ is the robot base-frame rotation. $p^{\text{robot}} \in \mathbb{R}^3$ is the robot base-frame position. q is the joint DOF positions (excluding wheels). q_{hl} is the hind-Leg DOF position. α is the $\angle($ robot $-x$ axis, world z axis$)$. f is the feet on the ground (binary). s is the standing robots (binary). It is worth noting that the reward function applied by this method [3] is only applicable to the bipedal standing task and needs to be redesigned for other movements.

However, our proposed method relies less on the intricacies of the reward function design, making it more accessible and easier to implement. The reward function in our framework consists of three main components: First, a reward for minimizing the deviation of the COM from a set value in the x-, y-, and z-directions. Second, a reward for minimizing the angle between the horizontal direction of the robot's coordinate system and the vertical direction of the world's coordinate system, encouraging the robot to maintain an upright posture during the standing task. Third, a penalty term, our reward function needs little change for the different tasks, only the addition of a penalty term for when the leg makes contact with the ground. The reward function can be formally defined as follows:

$$R = \gamma^t \cdot (r_{1gap} + r_{2gap} + r_{3gap} + r_{4gap}) + r_{angle} - C_{leg} \tag{13}$$

r_{gap} denotes the reward for tracking COM, which increases when the agent outputs an action that reduces the deviation of the robot's COM from the target position. For each agent, r_{gap} is the same. r_{range} is the posture reward item, which increases when the angle between the robot's trunk direction and the set target axis is reduced. For example, for the bipedal standing task, we set the Z-coordinate axis as the target, and when the agent outputs an action, the torso that remains vertical will receive a higher score, so this item can encourage the robot to maintain the standing posture. γ is the decay factor; the item is a hyperparameter, and the addition of the item and the value of the setting depends on experience and testing. After testing we found that adding a decay factor for the r_{gap} item was better, while r_{angle} was not suitable to add. t is the current time step, making the exponent γ^t decrease over time. By incorporating t in this way, the reward function can be dynamically adjusted, allowing the robot to focus on maintaining balance and stability initially, and then, gradually shifting attention to refining its movements and achieving more sophisticated locomotion patterns as the episode progresses. This temporal scaling helps in balancing the immediate and long-term objectives of the learning process. The C_{leg} is a penalty item with a negative reward that increases when the leg we set touches the ground, which in turn affects the global score, so this item makes the agent less likely to choose actions that we discourage. For example, in the bipedal standing task, we do not want the front left and front right legs to touch the ground, so we add penalties to both agents' actions. Here, we define

$$\begin{aligned} r_{gap} &= e^{-\left(|x_{com} - x_{target}| + |y_{com} - y_{target}| + |z_{com} - z_{target}|\right)} \\ r_{angle} &= e^{-|\psi|} \end{aligned} \tag{14}$$

where:

- x_{com}, y_{com}, and z_{com} represent the current position of the COM in the x-, y-, and z-directions, respectively. x_{target}, y_{target}, and z_{target} denote the target position of the COM in the x-, y-, and z-directions, respectively.

- ψ is the angle between the orientation of the robot trunk and the vertical direction of the world's coordinate system.
- γ is the decay factor for the leg contact penalty term. t represents the current time step.
- C_{leg} is a binary variable indicating whether the specified leg is in contact with the ground (1 for contact, 0 for no contact).

Our framework's simplicity and flexibility allow for the learning of advanced skills, such as bipedal walking for the ANYmal C robot, with minimal modifications to the reward function. To achieve this, we set the target trajectory of the COM to a desired height (e.g., $z_{target} = 1.5$) and specify the direction of motion (e.g., along the horizontal y-direction). By providing these straightforward guidelines, the robot can learn to walk in a bipedal manner without the need for a complex, task-specific reward function. Combining the multi-head agent structure, distributed multi-agent reinforcement learning, and the simplified reward function enables our framework to effectively learn complex behaviors while maintaining a high degree of generalization and adaptability to various tasks.

4. Experiments and Results

4.1. Experiment Setup

We used Isaac Gym, a simulation environment that allows large-scale parallel training using GPUs [26], and ANYmal C Robotics, developed by ANYRobotics [27], which is very powerful and has good support for RL-related algorithms, as our experimental platform. We incorporated accurate models of the robot's kinematics, inertial properties, and actuator characteristics based on the specifications provided by the manufacturer. Based on these data, our simulation environment was carefully designed and validated to closely mimic the physical properties, dynamics, and constraints of the real robot. For specific simulation parameters, the stiffness is 85.0 Nm/rad, damping is 2.0 Nm/rad, action scale is set as 0.5, control frequency is 18 Hz, gravity is -9.81 N, number of position iterations is 4, number of velocity iterations is 1, bounce threshold velocity is 4, and maximum depenetration velocity is 100.

To bridge the gap between simulation and display, we use domain randomization techniques that allow agents to cope with environments that vary over a range rather than a single fixed environment. Specific parameters include, initial position range 0 s$\sim$60 cm, initial orientation range $0\sim2\pi$ rad, joint static ratio errors ±0.08, joint dynamic ratio errors ±0.05, friction coefficient ±0.05, joint damping range 0.5$\sim$1.5 N, joint stiffness range 0.5$\sim$1.5 N, mass errors ±1.5 g, and height of COM errors ±0.5 cm.

We ran 4096 parallel simulations on an Nvidia RTX4070 GPU, and we learned three tasks, namely, walking on three legs, walking upright on two legs, and jumping on one leg. We observed and recorded experimental data. The videos of the simulation results for all experiments can be found in the Supplementary Materials with download links.

4.2. Training

To evaluate the effectiveness of our proposed method, we compared its performance with the popular reinforcement learning algorithm Proximal Policy Optimization (PPO). The results of this comparison are presented in Figure 4. The x-axis of the graph represents the number of samples used during training, while the y-axis indicates the episode length, which serves as a measure of the learned policy's effectiveness. Presenting the training results, particularly the rewards, in a unified graph is challenging due to the significant differences between single-agent reinforcement learning and the action space in our framework. These differences make it difficult to directly compare the reward values across different settings. However, to provide a meaningful comparison and insight into the effectiveness of the learned strategies, we analyzed the episode length during training. The episode length represents the duration for which the robot can successfully execute a learned strategy without failing and requiring a reset of the environment. A higher episode length indicates that the agent has learned a more robust and effective policy.

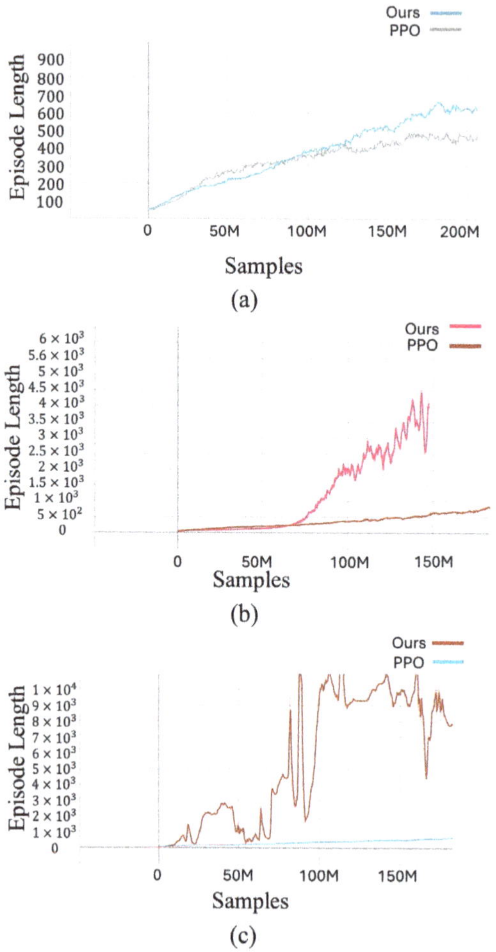

Figure 4. Comparison of the proposed method with PPO on different tasks: (**a**) Three-legged walking, (**b**) two-legged upright walking, and (**c**) one-legged walking.

In experiments, the environment is reset whenever the robot falls. Consequently, a higher episode length on the *y*-axis indicates a more effective learned policy, as it demonstrates the robot's ability to maintain stable walking for an extended period without falling. The steepness of the learning curve reflects the speed at which the policy is learned.

Our method achieved an advantage over PPO in each of the tasks investigated. Notably, we observed that the performance gap between the two methods widened as the difficulty of the task increased. In Figure 4a, the task involves walking using three legs. While PPO manages to learn a relatively effective policy, our method demonstrates a higher success rate and faster learning speed. Figure 4b presents the results for the task of walking upright using two legs, which poses a greater challenge. Initially, our method exhibits similar performance to PPO. However, after 70 M samples, the score of our method rapidly improves, eventually surpassing PPO in both training speed and final performance. Further observations revealed that PPO failed to learn an effective policy for walking upright on two legs. The most demanding task, one-legged walking, is depicted in Figure 4c. Maintaining an upright posture on a single leg is highly unstable due to the robot's structural constraints, resulting in significant fluctuations in our method's learning curve. Despite these challenges, our method ultimately learns an effective policy. In contrast, PPO once again fails to learn a successful policy for this task.

In terms of training results, our approach allows for the learning of highly dynamic locomotion or skills without complex reward design.

4.3. Simulation Results

In the simulation experiments, we observed the robot's action posture under various conditions. We first conducted a three-legged walking experiment, and the results are presented in Figure 5. In this experiment, we added a penalty term for one leg being in contact with the ground, set the center-of-mass height position to $z = 6$, and specified the motion along the x-axis. The robot's objective was to walk along the x-direction without falling. We observed that the robot could perform stable walking while lifting one leg. To investigate the effect of lifting different legs, we tested the robot's performance when lifting the back leg (Figure 5a) and the left front leg (Figure 5b). In both cases, the robot maintained stable walking, suggesting that all four of our agents learned effective policies to help stabilize the tracking of the center-of-mass position.

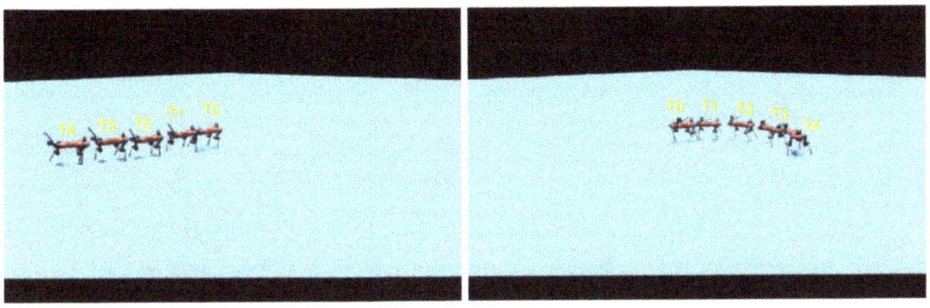

(**a**) Case 1 (**b**) Case 2

Figure 5. Three-legged walking experiment: (**a**) Lifting the back leg and (**b**) lifting the left front leg. The robot demonstrates stable walking while lifting one leg, indicating that all four agents learned effective policies to stabilize the tracking of the center-of-mass position.

We proceeded to test the bipedal walking experiment, and the results are shown in Figure 6. In this experiment, we set a penalty term for the left front and right front legs being in contact with the ground, adjusted the center-of-mass height position to $z = 15$, and specified movement along the x-axis. The robot's task was to walk along the x-direction without falling. We observed that the robot could perform stable walking using both legs. Interestingly, even though two legs were not actively involved in walking, their locomotion still contributed to stabilizing the position of the center of mass. The reduced weight of the locomotion of these two legs did not adversely affect the walking leg, thanks to the coordination achieved through MARL.

Finally, we conducted a single-leg walking experiment, which posed a significant challenge. The result is shown in Figure 7. In this experiment, we set a penalty term for three legs being in contact with the ground, adjusted the center-of-mass height position to $z = 14$, and specified movement along the x-axis. The robot's objective was to run along the x-direction without falling. We found that the robot could stabilize its balance on one leg, but maintaining the direction of motion proved to be a struggle. The robot managed to jump as far as possible in the set direction of motion but with a large deviation in direction.

These simulation results demonstrate the effectiveness of our proposed method in enabling the robot to learn stable walking policies under various conditions, including three-legged, bipedal, and single-leg walking. The multi-head agent structure and distributed multi-agent reinforcement learning approach allow the robot to coordinate its locomotion and maintain stability, even when faced with challenging tasks such as single-leg walking.

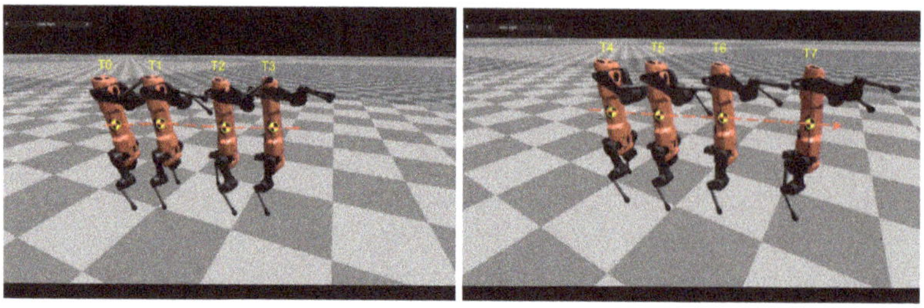

(**a**) Process 1 (**b**) Process 2

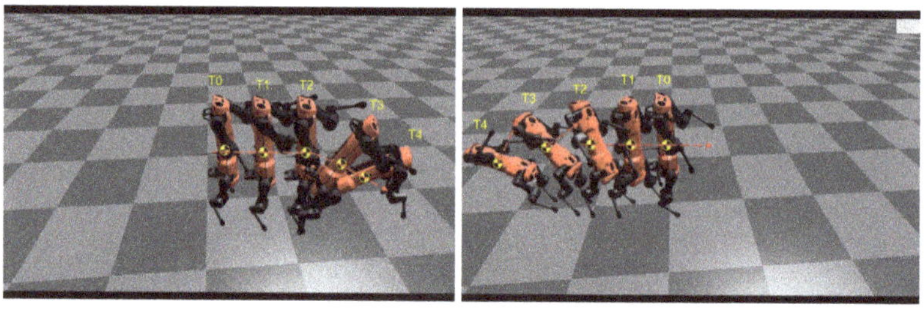

(**c**) Fail Case 1 (**d**) Fail Case 2

Figure 6. Bipedal walking experiment: (**a**,**b**) show the robot demonstrates stable walking using two legs, with the locomotion of the non-walking legs contributing to the stabilization of the center-of-mass position. (**c**,**d**) show two failure cases of the comparison method PPO.

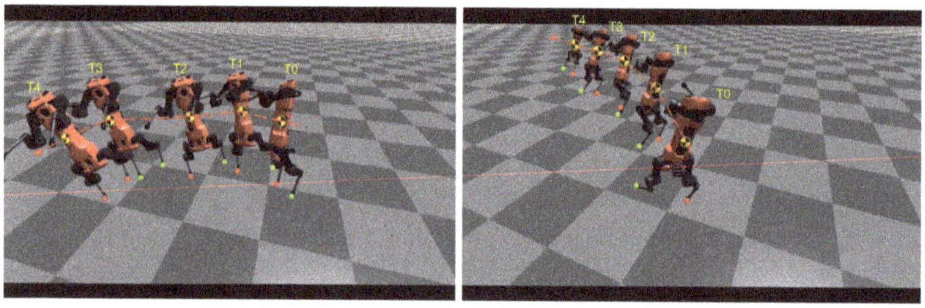

(**a**) Case 1 (**b**) Case 2

Figure 7. Single-leg walking experiment: The robot stabilizes its balance on one leg but struggles to maintain the direction of motion. Positions where the legs are in contact with the ground are marked by red markers, and positions where they are not in contact are marked by green markers. Jumping as far as possible in the set direction with a large deviation in direction.

We further performed a quantitative analysis to compare the stabilization of single-agent reinforcement learning with PPO, MAPPO, MPC. Our proposed framework combines RMP, a multi-head structure, and MARL. By including MAPPO, a state-of-the-art MARL algorithm, in our comparison, we aim to isolate and evaluate the effectiveness of the RMP and the multi-head structure components within our framework. Comparing our framework's performance against MAPPO allows us to assess the extent to which RMP and the multi-head structure contribute to the overall effectiveness of our approach, beyond the benefits provided by MARL alone. Thus, the MAPPO method has the same parameter settings as our method except that it does not have the RMP and polytope structure.

The traditional approach in MPC was used; this approach learns the quadrupedal robot bipedal walking task by many repetitions [28]. By comparing our framework against MPC, we aim to demonstrate the advantages of our learning-based approach in terms of adaptability and robustness. The dynamic equations of motion for such a robot can be derived from the Lagrangian, which is the difference between the kinetic energy (T) and potential energy (V) of the system, $L = T - V$, where T is the kinetic energy of the robot and V is the potential energy of the robot. Lagrange's equations of motion are defined as $\frac{d}{dt}\left(\frac{\partial L}{\partial \dot{q}_j}\right) - \frac{\partial L}{\partial q_j} = \tau_j$, where q_j is the generalized coordinates (positions of joints), $\dot{q}_j$ is the derivatives of the generalized coordinates (velocities), and τ_j is the generalized forces acting on the joints. To ensure stability, the center of mass must remain within the support polygon defined by the contact points of the feet on the ground. For a bipedal stance using legs 1 and 2, the constraint can be represented as $ax_{\text{COM}} + by_{\text{COM}} \leq c$, where a, b, and c are coefficients calculated based on the positions of the feet (x_1, y_1) and (x_2, y_2). Each joint angle q_j and torque τ_j should respect the mechanical limits of the robot: $q_j^{\min} \leq q_j \leq q_j^{\max}$, $|\tau_j| \leq \tau_j^{\max}$.

The cost function is designed to balance the objectives of tracking a desired stance, minimizing control effort, and penalizing deviations from constraints. It is defined as follows: $= \sum_{k=0}^{N-1} \left(w_p \|x_k - x_{\text{ref}}\|^2 + w_u \|u_k\|^2 + w_c c_k\right)$. Where x_k is the state vector at step k, including joint angles and velocities. x_{ref} is the reference state vector, representing the desired stance. u_k is the control input (torques) at step k. c_k is the penalty for constraint violations at step k, which could be based on balance and physical limits. w_p, w_u, and w_c are the weights balancing the importance of pose tracking, control effort, and constraint adherence, respectively. The full results are shown in Table 1.

Table 1. Comparison of stability and success rates of different methods under various tasks.

Task	Method	X Error	Y Error	Z Error	Success Rate
4-Leg Walking	Ours	±0.04	±0.04	±0.26	99.6%
	PPO	±0.04	±0.06	±0.38	98.2%
	MAPPO	±0.07	±0.09	±0.33	99.6%
	MPC	±0.02	±0.03	±0.24	100%
3-Leg Walking	Ours	±0.05	±0.04	±0.36	98.6%
	PPO	±0.06	±0.5	±0.58	94.6%
	MAPPO	±0.17	±0.14	±0.63	96.5%
	MPC	±0.04	±0.03	±0.44	98.3%
2-Leg Walking	Ours	±1.15	±1.13	±0.62	98.3%
	PPO	±6.4	±5.7	±3.5	42.7%
	MAPPO	±4.8	±4.5	±1.44	83.5%
	MPC	±1.07	±1.32	±0.82	98.3%
1-Leg Walking	Ours	±5.8	±7.2	±1.82	86.2%
	PPO	±17.4	±14.2	±6.7	*Failed*
	MAPPO	±10.2	±13.7	±5.8	*Failed*
	MPC	±6.3	±10.3	±4.6	64.2%

To evaluate the performance of our proposed method and compare it with other state-of-the-art approaches, we conducted a quantitative analysis of the experimental results. For each task, we measured the average range of deviation of the robot's center of mass from the target in the x-, y-, and z-directions, which reflects the degree of stabilization of the movement. Additionally, we calculated the success rate of stabilized walking over 300 simulation experiments in non-rendering mode.

Although the focus of this work was not on the 4-leg walking task, it is still informative as basic locomotion for quadrupedal robots, so we present the results here. This is implemented by not adding any penalty term for leg–ground contact. It should be noted that our approach is more suitable for highly dynamic and unstable situations. Due to the structure of the quadruped robot, having four legs touching the ground is already a relatively stable configuration, so the 4-legged walking task does not show a significant advantage over other approaches. Taken together, all methods achieved very high success rates as well as COM tracking accuracy. Due to the lack of fine-tuning for the reward function, or the introduction of imitation learning, our motion, while ensuring stability, did not exhibit the same natural movement as seen in other outstanding works. Nonetheless, the advantage of scalability in our approach was still confirmed.

In the 3-legged walking task, all methods performed relatively well. Although there was a large deviation in the z-direction due to the walking ups and downs associated with the three legs, it was within normal limits. As the tasks became more challenging, the performance of the compared methods began to deteriorate.

In the 2-legged walking task, the unstable posture led to an increased deviation range for all methods. However, our proposed approach still maintained deviations within a controllable range and demonstrated a clear advantage over PPO and MAPPO. The success rate of PPO decreased significantly in this task, with many instances of unstable walking. In contrast, MPC maintained a very good result.

The 1-legged walking task posed the greatest challenge, as the robot's posture became extremely unstable. Consequently, the deviations from all methods became very large, and it was difficult for the robot to maintain the direction of motion. Despite these challenges, our method showed promising results, with a relatively small deviation in the z-direction, indicating that it could still maintain the standing posture. The success rate also reflected that our method ensured the motion of the robot, while PPO and MAPPO were unable to control the robot in this stance. The success rate of MPC dropped drastically, revealing its inability to fully control the robot's locomotion and maintain stability.

Our proposed method maintains stable and controlled movement across all tasks, particularly in the challenging 2-legged and 1-legged walking scenarios. The combination of multi-agent reinforcement learning and dynamic space decomposition enables our approach to effectively coordinate and adapt the robot's locomotion, resulting in higher success rates and better stabilization compared to state-of-the-art methods.

5. Discussion

5.1. Conclusions

We proposed a method enabling quadrupedal robots to learn advanced and unstable motions, such as walking on one, two, or three legs. Traditional control methods struggle with such tasks due to instability and the need for precise leg coordination. Our proposed approach offers a promising solution for enabling quadrupedal robots to learn complex and adaptive locomotion behaviors. The combination of multi-agent reinforcement learning, RMPs, and dynamic space decomposition allows for the effective coordination and control of the robot's legs and stable and efficient movement, even in highly dynamic and unstable situations, by tracking COM alone.

Our architecture uses a distributed multi-agent framework for training coordinated locomotion policies, a structural multi-head agent for generating RMPs for single-leg control and center-of-mass tracking, and a bottom layer for action execution to maintain safety and stability. By treating each leg as an independent agent with a fully actuated subsystem, we effectively apply RMPs to underactuated motion control. In contrast to traditional RL methods, our approach can be used for many different tasks with only penalty item changes to the reward function, making it highly generalizable.

The experimental results show our method outperforms state-of-the-art reinforcement learning algorithms like PPO and MAPPO, as well as traditional MPC, particularly in challenging bipedal and one-leg tasks. Our method learns useful policies very quickly, even if the rewards are sparse. Our approach resulted in lower COM deviations and higher success rates, demonstrating its ability to maintain stable, controlled movement across all tasks.

It is important to note that 4-legged walking is not discussed in this paper, because it has been extensively studied, with mature solutions available. Our primary objective in this work is to showcase the capabilities and advantages of our proposed control method in enabling advanced locomotion skills, include versatility, robustness, and ability to adapt to extreme and unconventional locomotion scenarios. Our method can also be easily used in conjunction with other quadrupedal walking methods if necessary. Therefore, it will not be further elaborated on.

5.2. Limitations and Future Work

Despite the promising results, our proposed method has several limitations. Currently, we are unable to experiment on real robots due to the simulation-to-reality gap and safety concerns that need to be verified. Additionally, the transformation process from quadrupedal to bipedal morphology is not yet possible with our method, which may require further training with the aid of imitative learning. Moreover, our experiments are limited to flat terrain, and we need to introduce more environments with terrain and obstacles to train the robot to operate in the real world.

Theoretically, our approach could be applied to various robotic platforms, such as humanoid robots, wheeled–legged robots, or other robots with multiple degrees of freedom. However, applying our approach to robots with different structures and control requirements may necessitate additional considerations and modifications. While our approach provides a general framework for controlling multi-degree-of-freedom robots, developing a unified framework that can seamlessly control different robot structures without secondary training remains a challenge.

Future work could focus on addressing these limitations by extending the approach to more diverse terrains and environments and investigating the application of transfer learning to accelerate the learning process for new tasks. Furthermore, combining the proposed method with advanced generative AI and learning from large amounts of video data could enable robots to learn more complex motor skills and interaction skills by using AI to plan more complex COM motion trajectories [29].

Supplementary Materials: The following supporting information can be downloaded at: https://www.mdpi.com/article/10.3390/robotics13060086/s1, Video S1: Simulation Experiment Results.

Author Contributions: Conceptualization, Y.W.; data curation, Y.W.; formal analysis, Y.W.; funding acquisition, Y.W., R.S. and Y.Y.; investigation, R.S. and Y.Y.; methodology, Y.W.; software, Y.W.; supervision, R.S. and Y.Y.; validation, Y.W.; writing—original draft, Y.W.; writing—review and editing, R.S. and Y.Y. All authors have read and agreed to the published version of the manuscript.

Funding: This work was supported by JST SPRING, Grant Number JPMJSP2124 (Japan Science and Technology Agency Support for Pioneering Research Initiated by the Next Generation), JSPS JP22H00545, JP22H05002 (Japan Society for the Promotion of Science) and NEDO JPNP20006 (New Energy and Industrial Technology Development Organization) in Japan.

Data Availability Statement: Data are contained within the article.

Conflicts of Interest: The remaining authors declare that the research was conducted in the absence of any commercial or financial relationships that could be construed as potential conflicts of interest.

References

1. Bjelonic, M.; Grandia, R.; Harley, O.; Galliard, C.; Zimmermann, S.; Hutter, M. Whole-body MPC and online gait sequence generation for wheeled-legged robots. In Proceedings of the 2021 IEEE/RSJ International Conference on Intelligent Robots and Systems (IROS), Prague, Czech Republic, 27 September–1 October 2021; pp. 8388–8395.
2. Nagano, K.; Fujimoto, Y. Zero Moment Point Estimation Based on Resonant Frequencies of Wheel Joint for Wheel-Legged Mobile Robot. *IEEJ J. Ind. Appl.* **2022**, *11*, 408–418. [CrossRef]
3. Smith, L.; Kew, J.C.; Li, T.; Luu, L.; Peng, X.B.; Ha, S.; Tan, J.; Levine, S. Learning and adapting agile locomotion skills by transferring experience. *arXiv* **2023**, arXiv:2304.09834.
4. Qi, J.; Gao, H.; Su, H.; Han, L.; Su, B.; Huo, M.; Yu, H.; Deng, Z. Reinforcement learning-based stable jump control method for asteroid-exploration quadruped robots. *Aerosp. Sci. Technol.* **2023**, *142*, 108689. [CrossRef]
5. Tang, Z.; Kim, D.; Ha, S. Learning agile motor skills on quadrupedal robots using curriculum learning. In *Proceedings of the 3rd International Conference on Robot Intelligence Technology and Applications*; Springer: Singapore, 2021; Volume 3.
6. Vollenweider, E.; Bjelonic, M.; Klemm, V.; Rudin, N.; Lee, J.; Hutter, M. Advanced skills through multiple adversarial motion priors in reinforcement learning. In Proceedings of the 2023 IEEE International Conference on Robotics and Automation (ICRA), London, UK, 29 May–2 June 2023; IEEE: Piscataway, NJ, USA, 2023.

7. Mattamala, M.; Chebrolu, N.; Fallon, M. An efficient locally reactive controller for safe navigation in visual teach and repeat missions. *IEEE Robot. Autom. Lett.* **2022**, *7*, 2353–2360. [CrossRef]
8. Ratliff, N.D.; Issac, J. Riemannian motion policies. *arXiv* **2018**, arXiv:1801.02854.
9. Shaw, S.; Abbatematteo, B.; Konidaris, G. RMPs for safe impedance control in contact-rich manipulation. In Proceedings of the 2022 International Conference on Robotics and Automation (ICRA), Philadelphia, PA, USA, 23–27 May 2022; pp. 2707–2713.
10. Wingo, B.; Cheng, C.A.; Murtaza, M.; Zafar, M.; Hutchinson, S. Extending Riemmanian motion policies to a class of underactuated wheeled-inverted-pendulum robots. In Proceedings of the 2020 IEEE International Conference on Robotics and Automation (ICRA), Paris, France, 31 May–31 August 2020; IEEE: Piscataway, NJ, USA, 2020.
11. Marew, D.; Lvovsky, M.; Yu, S.; Sessions, S.; Kim, D. Integration of Riemannian Motion Policy with Whole-Body Control for Collision-Free Legged Locomotion. In Proceedings of the 2023 IEEE-RAS 22nd International Conference on Humanoid Robots (Humanoids), Austin, TX, USA, 12–14 December 2023.
12. Cheng, C.-A.; Mukadam, M.; Issac, J.; Birchfield, S.; Fox, D.; Boots, B.; Ratliff, N. Rmpflow: A computational graph for automatic motion policy generation. In *Algorithmic Foundations of Robotics XIII: Proceedings of the 13th Workshop on the Algorithmic Foundations of Robotics*; Springer International Publishing: Cham, Switzerland, 2020; pp. 441–458.
13. Bellegarda, G.; Chen, Y.; Liu, Z.; Nguyen, Q. Robust high-speed running for quadruped robots via deep reinforcement learning. In Proceedings of the 2022 IEEE/RSJ International Conference on Intelligent Robots and Systems (IROS), Kyoto, Japan, 23–27 October 2022; pp. 10364–10370.
14. Ladosz, P.; Weng, L.; Kim, M.; Oh, H. Exploration in deep reinforcement learning: A survey. *Inf. Fusion* **2022**, *85*, 1–22. [CrossRef]
15. Mnih, V.; Kavukcuoglu, K.; Silver, D.; Rusu, A.A.; Veness, J.; Bellemare, M.G.; Graves, A.; Riedmiller, M.; Fidjeland, A.K.; Ostrovski, G.; et al. Human-level control through deep reinforcement learning. *Nature* **2015**, *518*, 529–533. [CrossRef] [PubMed]
16. Tan, H. Reinforcement learning with deep deterministic policy gradient. In Proceedings of the 2021 International Conference on Artificial Intelligence, Big Data and Algorithms (CAIBDA), Sanya, China, 23–25 April 2021; pp. 82–85.
17. Haarnoja, T.; Zhou, A.; Abbeel, P.; Levine, S. Soft actor-critic: Off-policy maximum entropy deep reinforcement learning with a stochastic actor. In Proceedings of the International Conference on Machine Learning (ICML), Stockholm, Sweden, 10–15 July 2018; pp. 1861–1870.
18. Zhang, K.; Yang, Z.; Başar, T. Multi-agent reinforcement learning: A selective overview of theories and algorithms. In *Handbook of Reinforcement Learning and Control*; Springer: Cham, Switzerland, 2021; pp. 321–384.
19. Canese, L.; Cardarilli, G.C.; Di Nunzio, L.; Fazzolari, R.; Giardino, D.; Re, M.; Spanò, S. Multi-agent reinforcement learning: A review of challenges and applications. *Appl. Sci.* **2021**, *11*, 4948. [CrossRef]
20. Wang, Y.; Sagawa, R.; Yoshiyasu, Y. A Hierarchical Robot Learning Framework for Manipulator Reactive Motion Generation via Multi-agent Reinforcement Learning and Riemannian Motion Policies. *IEEE Access* **2023**, *1*, 126979–126994. [CrossRef]
21. Liu, M.; Qu, D.; Xu, F.; Zou, F.; Di, P.; Tang, C. Quadrupedal robots whole-body motion control based on centroidal momentum dynamics. *Appl. Sci.* **2019**, *9*, 1335. [CrossRef]
22. Luo, J.; Li, C.; Fan, Q.; Liu, Y. A graph convolutional encoder and multi-head attention decoder network for TSP via reinforcement learning. *Eng. Appl. Artif. Intell.* **2022**, *112*, 104848. [CrossRef]
23. Yu, C.; Velu, A.; Vinitsky, E.; Gao, J.; Wang, Y.; Bayen, A.; Wu, Y. The surprising effectiveness of PPO in cooperative multi-agent games. In *Advances in Neural Information Processing Systems*; Curran Associates, Inc.: Red Hook, NY, USA, 2022; Volume 35, pp. 24611–24624.
24. Lowe, R.; Wu, Y.I.; Tamar, A.; Harb, J.; Abbeel, O.P.; Mordatch, I. Multi-agent actor-critic for mixed cooperative-competitive environments. In *Advances in Neural Information Processing Systems 30*; Curran Associates, Inc.: Red Hook, NY, USA, 2017.
25. Schulman, J.; Wolski, F.; Dhariwal, P.; Radford, A.; Klimov, O. Proximal policy optimization algorithms. *arXiv* **2017**, arXiv:1707.06347.
26. Makoviychuk, V.; Wawrzyniak, L.; Guo, Y.; Lu, M.; Storey, K.; Macklin, M.; Hoeller, D.; Rudin, N.; Allshire, A.; Handa, A.; et al. Isaac Gym: High performance GPU-based physics simulation for robot learning. *arXiv* **2021**, arXiv:2108.10470.
27. ANYmal C—The Next Step in Robotic Industrial Inspection. Available online: https://www.anybotics.com/news/the-next-step-in-robotic-industrial-inspection (accessed on 20 August 2019).
28. Corbères, T.; Flayols, T.; Léziart, P.A.; Budhiraja, R.; Souères, P.; Saurel, G.; Mansard, N. Comparison of predictive controllers for locomotion and balance recovery of quadruped robots. In Proceedings of the 2021 IEEE International Conference on Robotics and Automation (ICRA), Xi'an, China, 30 May–5 June 2021; pp. 5021–5027.
29. Brohan, A.; Brown, N.; Carbajal, J.; Chebotar, Y.; Chen, X.; Choromanski, K.; Ding, T.; Driess, D.; Dubey, A.; Finn, C.; et al. RT-2: Vision-Language-Action Models Transfer Web Knowledge to Robotic Control. *arXiv* **2023**, arXiv:2307.15818.

robotics

MDPI

Article

Learning to Walk with Adaptive Feet

Antonello Scaldaferri [1,2,*], Franco Angelini [1,2] and Manolo Garabini [1,2]

[1] Centro di Ricerca "Enrico Piaggio", Università di Pisa, Largo Lucio Lazzarino 1, 56122 Pisa, Italy; franco.angelini@unipi.it (F.A.); manolo.garabini@unipi.it (M.G.)
[2] Dipartimento di Ingegneria dell'Informazione, Università di Pisa, Via Girolamo Caruso 16, 56122 Pisa, Italy
* Correspondence: antonello.scaldaferri@phd.unipi.it; Tel.: +39-349-4908051

Abstract: In recent years, tasks regarding autonomous mobility favoredthe use of legged robots rather than wheeled ones thanks to their higher mobility on rough and uneven terrains. This comes at the cost of more complex motion planners and controllers to ensure robot stability and balance. However, in the case of quadrupedal robots, balancing is simpler than it is for bipeds thanks to their larger support polygons. Until a few years ago, most scientists and engineers addressed the quadrupedal locomotion problem with model-based approaches, which require a great deal of modeling expertise. A new trend is the use of data-driven methods, which seem to be quite promising and have shown great results. These methods do not require any modeling effort, but they suffer from computational limitations dictated by the hardware resources used. However, only the design phase of these algorithms requires large computing resources (controller training); their execution in the operational phase (deployment), takes place in real time on common processors. Moreover, adaptive feet capable of sensing terrain profile information have been designed and have shown great performance. Still, no dynamic locomotion control method has been specifically designed to leverage the advantages and supplementary information provided by this type of adaptive feet. In this work, we investigate the use and evaluate the performance of different end-to-end control policies trained via reinforcement learning algorithms specifically designed and trained to work on quadrupedal robots equipped with passive adaptive feet for their dynamic locomotion control over a diverse set of terrains. We examine how the addition of the haptic perception of the terrain affects the locomotion performance.

Keywords: dynamiclocomotion control; end-to-end learning; mechanically adaptive feet; quadrupedal robots; reinforcement learning

1. Introduction

Legged robots have emerged as a fascinating field of study and innovation in the realm of robotics. The unique capabilities of legged robots stem from the natural motion patterns of animals, which have evolved over millions of years to be efficient and effective in various landscapes. Drawing inspiration from nature's most agile creatures, quadrupedal robots aim to replicate the locomotion abilities of four-legged animals [1]. Their versatile and adaptive movement in unstructured environments has garnered significant interest across various industries, ranging from search-and-rescue operations [2] to the exploration of harsh terrains [3,4]. Legged locomotion holds a distinct advantage over wheeled or tracked locomotion as it allows robots to traverse challenging terrains, bypass obstacles, and maintain stability even in unpredictable environments [5]. Quadrupedal robots, in particular, have shown great promise due to their ability to distribute weight across four legs, ensuring better stability and load-bearing capacity.

Considerable effort and focus have gone into the design of robotic feet for stable interaction with the environment [6–9]. For instance, the authors in [10] developed passive mechanically adaptive feet for quadrupedal robots capable of adapting to different terrain surfaces, significantly reducing robot slippage with respect to flat and ball feet.

From the motion point of view, dynamic locomotion plays a pivotal role in achieving agile and efficient movements for legged robots [11]. Unlike static locomotion, dynamic locomotion emphasizes the agility of these machines. The exploration of dynamic locomotion control for quadrupedal robots has become a critical area of research, seeking to enhance their agility, stability, and overall performance. Through dynamic locomotion control, robots can adjust their gait, step frequency, and leg movements in response to varying terrains, slopes, or disturbances. This adaptability allows quadrupedal robots to quickly handle unexpected obstacles and maintain balance and stability while moving at different speeds. The development of effective dynamic locomotion control algorithms presents numerous challenges [12]. Engineers and researchers must face complex issues such as environment reconstruction, state estimation, motion planning, and feedback control. Furthermore, achieving seamless coordination between multiple legs and ensuring smooth transitions between different gaits demands sophisticated control strategies.

Recent works have demonstrated that by exploiting highly parallel GPU-accelerated simulators and deep reinforcement learning, end-to-end policies can be trained in minutes even for complex control tasks such as quadrupedal robot dynamic locomotion [13,14].

To the best of the authors' knowledge, there is no previous research addressing the problem of designing dynamic locomotion control algorithms that work on quadrupedal robots equipped with the passive adaptive feet presented in [10] (Figure 1), which consistently differ from the well-known and extensively investigated ball and flat feet. Therefore, the motivation of this work is the lack of control approaches designed and thought to fully exploit and leverage the information provided by this type of adaptive feet.

Figure 1. Quadrupedal robot dynamic locomotion with adaptive feet.

Given the kinematic and dynamic high complexity of these adaptive feet and considering the lack of good informative mathematical models regarding dynamic locomotion in harsh environments, as in the context of robotic environmental monitoring of natural habitats, we decided to address this problem by leveraging data-driven and learning-based methods over model-based approaches.

The contribution of this work is the design, training, and investigation of different learning-based end-to-end control policies enabling quadrupedal robots equipped with the passive adaptive feet presented in [10] to perform dynamic locomotion over a diverse set of terrains. For end-to-end policies, we mean that between the input and the output of the controller, there is only a single neural network without any additional model-based sub-module. The adaptive feet used in this work come from previous research [10]; thus, they are not a contribution of this work.

This paper is organized as follows: in Section 2, we present the state-of-the-art learning-based quadrupedal robot dynamic locomotion control approaches and other related works of the problem faced; in Section 3, we present the method used to solve the quadrupedal robot dynamic locomotion control problem with adaptive feet; in Section 4, the results obtained in simulation are shown; in Section 5, we compare the results of the different trained control policies, and we discuss them; and in Section 6, we conclude and present potential future work that could improve the performance of the system.

2. Related Work

State-of-the-art learning-based quadrupedal robot dynamic locomotion control approaches use hybrid methods like [15], where a policy is trained to provide a few parameters that modulate model-based foot trajectory generators. In [16], the authors train in simulation, using a hybrid simulator similar to the one in [17], a non-exteroceptive student policy to imitate the behavior of a teacher policy that has access to privileged information not available to the real robot, and using a temporal convolutional neural network-based policy [18]. In [19], the authors similarly train a LiDAR- and RGBD-based perceptive policy capable of distinguishing when to rely on proprioception rather than exteroception using a recurrent neural network-based belief-state encoder.

These methods are hybrid in the sense of [15], and they are not end-to-end in the sense that they do not rely on a purely learning-based approach. End-to-end pipelines have the freedom to learn behaviors that may not be possible to perform with an a priori fixed structure. On the other hand, hybrid approaches provide a better warm start in the learning phase, and they are more reliable because of their higher level of interpretability and robust in terms of stability because of their use of well-known first principles. However, certificated learning strategies [20] can be used to verify and satisfy safety and stability requirements directly during the training process. Figure 2 highlights the main differences in terms of different controller components of hybrid architectures versus end-to-end ones.

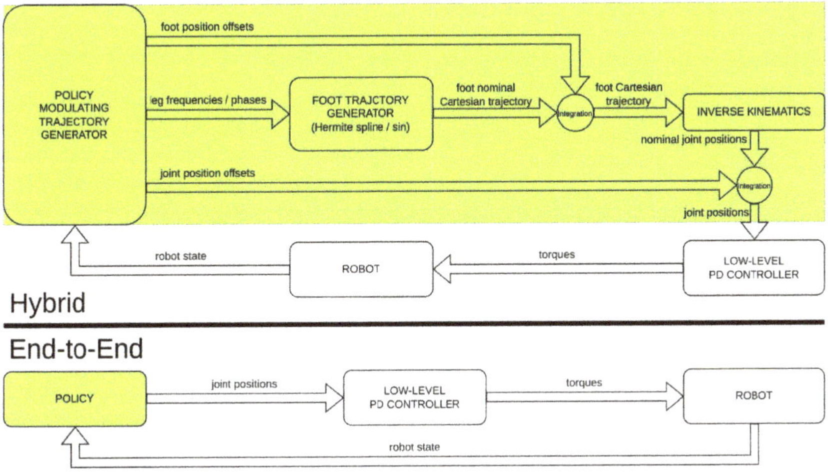

Figure 2. Differences between hybrid [15] and end-to-end [14] approaches.

More recently, end-to-end methods [14] instead have been demonstrated to be able to successfully learn perceptive policies, whose performance is comparable to that of hybrid methods [19], without any model-based trajectory generator or other prior knowledge and without any privileged learning or teacher–student-based imitation learning. Other works [21–25] have been carried out to showcase how using end-to-end learning-based pipelines enables advanced motor skills for quadrupedal robot locomotion and navigation in unstructured and challenging environments. These methods, however, have all been designed to run on robots equipped with standard ball feet, with point-based contact dynamics assumptions.

Passive adaptive feet [10] have been designed to be able to adapt to the shape of any kind of obstacle, maximizing grip and avoiding slippage. Moreover, they are equipped with a sensing system capable of providing information about the environment and the foot interaction with the surroundings. The authors in [26] developed a foothold selection algorithm for quadrupedal robots equipped with SoftFoot-Q [10] mechanically adaptive soft feet. As in [27], on the baseline [28], a dataset of elevation maps of synthetically generated terrains is acquired and collected in simulation. Then a polynomial fitting algorithm is used, along with other computations, to compute the associated cost maps, which are then used for the foothold selection. Since the terrain assessment and constraint checking are computationally expensive, a convolutional neural network (CNN), more precisely an efficient residual factorized convolutional neural network (ERFNet) [29], is trained and used to predict a good approximation of the cost map, which is then used to compute the potential footholds on the elevation map.

Another interesting work addressing real-time dynamic locomotion on rough terrains that deals with dynamic footholds is presented in [30], but it is still designed to run on robots equipped with standard ball feet. In [31], the authors aim to improve dynamic quadrupedal locomotion robustness by employing a fast model-predictive foothold-planning algorithm and an LQR controller applied to projected inverse dynamic control for robust motion tracking. The method shows robustness by being able to achieve good dynamic locomotion performance also with unmodeled adaptive feet, but it does not make any use of the supplementary haptic information of the terrain provided by these feet and thus does not exploit the benefits of their passive adaptability.

Despite the good work performed on footstep planning with the adaptive feet presented in [10], there is no work on dynamic locomotion control explicitly designed for them. All the mentioned methods employ static walking gait patterns, such as crawling, where robot speed is limited. Moreover, all the works presented so far address the problem of dynamic locomotion but not with the use of the adaptive feet presented in [10].

3. Method

We model the quadrupedal robot dynamic locomotion control problem with adaptive feet as a Markov decision process (MDP). The latter is a mathematical framework to formulate discrete-time decision-making processes. The MDP is the standard mathematical formalism when dealing with reinforcement learning problems. An MDP is described by a tuple of five elements $\langle \mathcal{S}, \mathcal{A}, \mathcal{R}, \mathcal{P}, \subset_0 \rangle$. $\mathcal{S}$ defines the state space, $\mathcal{A}$ defines the action space, $\mathcal{R}(s_t, a_t, s_{t+1}) : \mathcal{S} \times \mathcal{A} \times \mathcal{S} \rightarrow \mathbb{R}$ is the reward function, $\mathcal{P}(s_{t+1}|s_t, a_t)$ is the transition probability density function, and $\subset_0$ is the starting state distribution.

At each time step t, the learning agent takes an action $a_t \in \mathcal{A}$ based on the environment state $s_t \in \mathcal{S}$. The action is drawn from a parameterized policy $a_t \sim \pi_\theta(\cdot|s_t)$, with θ being the set of policy parameters. This results in a reward value r_t. In the case of policy optimization algorithms, among reinforcement learning methods, the goal is to find the optimal set of policy parameters θ^*, which maximizes the expected return $\mathcal{J}(\pi_\theta) = \int_\tau \mathcal{P}(\tau|\pi_\theta)\mathrm{R}(\tau) = \mathbb{E}_{\tau\sim\pi_\theta}[\mathrm{R}(\tau)]$, i.e., $\theta^* = \mathrm{argmax}_\theta \mathcal{J}(\pi_\theta)$. This goal is achieved through interactions with the environment in a trial-and-error fashion, where $\tau = (s_0, a_0, s_1, a_1, ...)$ are trajectories (sequences of states and actions) drawn from the policy, and $\mathrm{R}(\tau) = \sum_{t=0}^{\infty} \gamma^t r_t$ is the infinite-horizon discounted return (the cumulative reward), with discount factor $\gamma \in (0, 1)$ and r_t being the reward at time step t.

A Gaussian distribution is used to model the policy $\pi_\theta(a_t|s_t)$, i.e., $a_t \sim \mathcal{N}(\mu_\theta(s_t), \sigma_\theta)$. We use a multi-layer perceptron (MLP) for the mean value $\mu_\theta(\cdot)$ and a state-independent standard deviation σ_θ. The policy optimization is obtained through the proximal policy optimization (PPO) algorithm [32] from RL-Games [33].

We define our approach as end-to-end because it uses a single neural network $\mu_\theta(\cdot)$ as the only component of the controller.

As the authors in [26] highlight, the simulation of the adaptive foot [10] is slow because of its mechanical complexity, since it counts a total of 34 joints, and a quadrupedal robot employs four of these feet. Even using a GPU-accelerated physics-based simulator such as Isaac Sim [34], the simulation remains slow, especially considering that to exploit its high level of parallelism, thousands of robots are simultaneously simulated, which is useful to find optimal policies for reinforcement learning problems in a short time [13]. Moreover, even if Isaac Sim supports closed kinematic chain simulation, given the high complexity of the adaptive feet, the simulation of the closed kinematic model of the foot is highly unstable and often leads to non-physical behaviors when high contact forces are involved. For these reasons, we decided to employ two simplified approximations of the adaptive foot model shown in Figure 3a, both of which have an open kinematic chain structure, preserving the major degrees of freedom of their adaptability.

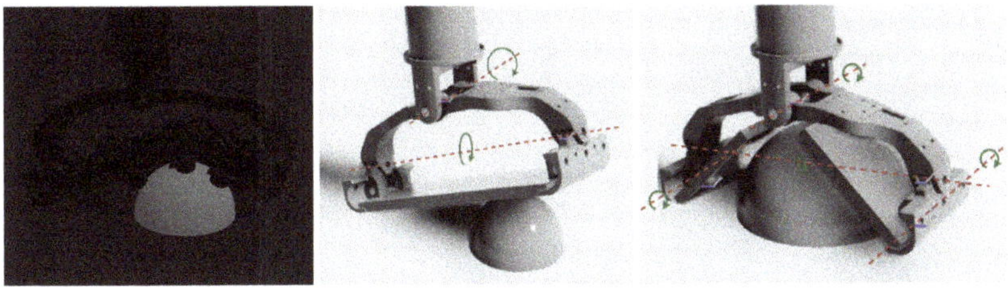

(**a**) (**b**) (**c**)

Figure 3. Proposed approximations of the SoftFoot-Q [10] (**a**) model: (**b**) Adaptive Flat Foot (AFF) and (**c**) Adaptive Open Foot (AOF). The passive DoF rotation axes are highlighted in red.

These model approximations are shown in Figure 3. In particular, the two simplified models that were designed are the Adaptive Flat Foot (AFF) model and the Adaptive Open Foot (AOF) model, represented in Figure 3b,c, respectively. The Adaptive Flat Foot keeps the pitch and roll joints of the ankle of the SoftFoot-Q shown in Figure 3a, with the three chains being replaced by a single rigid flat body for the sole (2 passive degrees of freedom in total). The Adaptive Open Foot instead splits the single rigid flat body of the sole into two bodies, each one having an additional rotational degree of freedom with regard to the transversal axis of the foot roll link, allowing the foot to adapt to the shapes of different terrain obstacles and artifacts (4 passive degrees of freedom in total).

The reinforcement learning environments, shown in Figure 4, used for training the dynamic locomotion control policies are based on the baselines available in [35], a collection of reinforcement learning environments for common and complex continuous-time control problems, implemented to run on Isaac Sim. They are the flat terrain locomotion environment, shown in Figure 4a, and the rough terrain locomotion environment, shown in Figure 4b. Both of them use the ANYmal quadrupedal robot platform [36], equipped with a total of 12 actuated degrees of freedom, 3 per leg, namely, Hip Abduction–Adduction (HAA), Hip Flexion–Extension (HFE), and Knee Flexion–Extension (KFE).

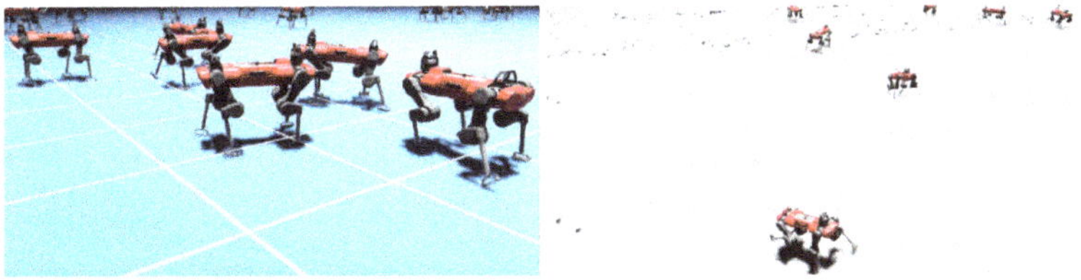

(**a**) Flat terrain. (**b**) Rough terrains.

Figure 4. Reinforcement learning environments.

As in [14], the rough terrain policies were trained in a game-inspired automatic curriculum fashion [37], starting from simpler terrain samples and increasing in complexity once the control policies allow the robots to traverse them [38]. Therefore, the environment terrain is built as a single set of different terrains (smooth slope, rough slope, stairs, stepping stones) at different levels of difficulty, following the setup shown in Figure 5.

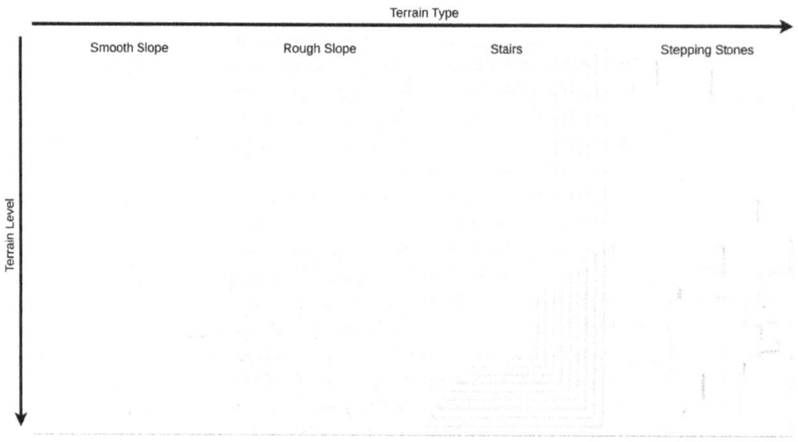

Figure 5. Terrain setup for the rough terrain environment.

For the flat terrain locomotion environment, the policy MLP $\mu_\theta(\cdot)$ has three hidden layers, with 256, 128, and 64 units, respectively, and a 12-dimensional output layer representing the 12 desired joint velocities $\dot{\boldsymbol{q}}_{\mathrm{d}}$, which are then integrated with the current joint positions $\boldsymbol{q}$, to obtain the desired joint positions $\boldsymbol{q}_{\mathrm{d}}$.

The rough terrain locomotion environment, instead, employs a policy MLP $\mu_\theta(\cdot)$ with three hidden layers with 512, 256, and 128 units, respectively, and a 12-dimensional output layer representing the 12 desired joint positions $\boldsymbol{q}_{\mathrm{d}}$ directly.

For both environments, a low-level PD controller is then used to compute the desired joint torques $\boldsymbol{\tau}_{\mathrm{d}}$ to apply to the robot actuated joints.

We considered different types of policy. In particular, the policy can either be perceptive, using a robot-centric elevation map of the surrounding terrain as shown in Figure 6 as part of the input, or blind, without considering it. Then, the policy can either exploit terrain haptic perception, taking as input also the joint positions and velocities of the feet, or not. Considering the combinations of these strategies, four types of locomotion policy have been defined: blind–non-haptic (B), perceptive–non-haptic (P), blind–haptic (BH), and perceptive–haptic (PH). The configuration of the policy observation space for each type of policy is given in Table 1. For the rough terrain locomotion task, all these types of policy were investigated, while for the flat terrain locomotion task, only the blind–non-haptic policy type (B) was considered.

The contact conditions of the adaptive feet with the terrain are not contained in any of the policy inputs. The feet information that the different policies take as input comprises the feet joint positions and velocities, as described in Table 1. Even if the feet joint positions and velocities cannot be directly measured via joint encoders, they can be easily estimated from the four IMU measurements available on each foot [10].

The input components of these different types of policy can be all or some of the following: base twist $\boldsymbol{v}^{\mathrm{b}}$, consisting of the 3 linear and angular velocities of the robot base; projected gravity $\boldsymbol{\phi}^{\mathrm{b}}_{\|g}$, that is, the projection of the gravity vector on the base frame axes, which gives information about the robot base orientation with regard to the gravity direction; command $\boldsymbol{v}^{\mathrm{b}}_{\mathrm{d}}$, that is, the user-provided desired robot base twist reference, whose details are given in the next paragraphs; joint positions $\boldsymbol{q}$, namely, the positions of the actuated joint; joint velocities $\dot{\boldsymbol{q}}$, namely, the velocities of the actuated joint; previous action $\boldsymbol{a}^{\mathrm{t}-1}$, that is, the output of the policy at the previous time step; height map $\boldsymbol{H}_{\mathrm{map}}$, i.e., the elevation map shown in Figure 6; feet joint positions $\boldsymbol{q}_{\mathrm{feet}}$, i.e., the positions of the feet passive joints; and feet joint velocities $\dot{\boldsymbol{q}}_{\mathrm{feet}}$, i.e., the velocities of the feet passive joints.

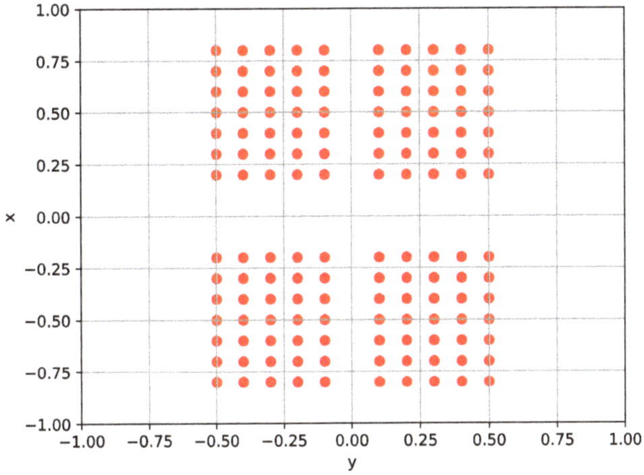

Figure 6. Robot-centric elevation map with sampled point coordinates.

Table 1. Policy observation space.

Input \ Policy Type	B	P	BH	PH
base twist $\boldsymbol{v}^{\mathrm{b}}$	✓	✓	✓	✓
projected gravity $\boldsymbol{\phi}^{\mathrm{b}}_{\parallel g}$	✓	✓	✓	✓
command $\boldsymbol{v}^{\mathrm{b}}_{\mathrm{d}}$	✓	✓	✓	✓
joint positions $\boldsymbol{q}$	✓	✓	✓	✓
joint velocities $\dot{\boldsymbol{q}}$	✓	✓	✓	✓
previous action $\boldsymbol{a}^{\mathrm{t}-1}$	✓	✓	✓	✓
height map $\boldsymbol{H}_{\mathrm{map}}$	✗	✓	✗	✓
feet joint positions $\boldsymbol{q}_{\mathrm{feet}}$	✗	✗	✓	✓
feet joint velocities $\dot{\boldsymbol{q}}_{\mathrm{feet}}$	✗	✗	✓	✓

The primary goal of the locomotion tasks is to track a user-provided commanded base twist reference $\boldsymbol{v}^{\mathrm{b}}_{\mathrm{d}}$ without falling. The command is composed of three components: a linear velocity along the robot sagittal axis ($v^{\mathrm{b_x}}_{\mathrm{d}}$), a linear velocity along the robot transversal axis ($v^{\mathrm{b_y}}_{\mathrm{d}}$), and an angular velocity around the robot longitudinal axis ($\omega^{\mathrm{b_z}}_{\mathrm{d}}$).

We also applied domain randomization [39], considering additive component-wise scaled random noise, generating randomized reference commands (the angular velocity command is generated starting from a randomized heading direction) and applying random disturbances to the robots through interspersed base random pushes.

The reward function is the sum of different conventional terms, and it aims to make the robot base twist:

$$\boldsymbol{v}^{\mathrm{b}} = \langle v^{\mathrm{b_x}}, v^{\mathrm{b_y}}, v^{\mathrm{b_z}}, \omega^{\mathrm{b_x}}, \omega^{\mathrm{b_y}}, \omega^{\mathrm{b_z}} \rangle, \tag{1}$$

to follow the commanded reference one

$$\boldsymbol{v}^{\mathrm{b}}_{\mathrm{d}} = \langle v^{\mathrm{b_x}}_{\mathrm{d}}, v^{\mathrm{b_y}}_{\mathrm{d}}, v^{\mathrm{b_z}}_{\mathrm{d}}, \omega^{\mathrm{b_x}}_{\mathrm{d}}, \omega^{\mathrm{b_y}}_{\mathrm{d}}, \omega^{\mathrm{b_z}}_{\mathrm{d}} \rangle, \tag{2}$$

with $v^{\mathrm{b_z}}_{\mathrm{d}}$, $\omega^{\mathrm{b_x}}_{\mathrm{d}}$, and $\omega^{\mathrm{b_y}}_{\mathrm{d}}$ always equal to zero.

The linear velocity term aims at minimizing the velocity tracking errors in the forward and lateral linear directions:

$$r_{v^{\mathrm{bxy}}} = e^{-\frac{\sum_{i\in(x,y)}(v_{\mathrm{d}}^{\mathrm{b}_i} - v^{\mathrm{b}_i})^2}{0.25}}. \quad (3)$$

The angular velocity term aims at minimizing the yaw angular velocity tracking error around the vertical axis of the robot base frame:

$$r_{\omega^{\mathrm{b_z}}} = \frac{e^{-\frac{(\omega_{\mathrm{d}}^{\mathrm{b_z}} - \omega^{\mathrm{b_z}})^2}{0.25}}}{2}. \quad (4)$$

The longitudinal linear velocity penalty term keeps the robot base's vertical velocity as small as possible to avoid jumping-like motions:

$$p_{v^{\mathrm{b_z}}} = C_{v^{\mathrm{b_z}}} {v^{\mathrm{b_z}}}^2. \quad (5)$$

The angular velocity penalty term keeps the robot base's roll and pitch angular velocities as small as possible to prevent the robot from rolling to the front or side:

$$p_{\omega^{\mathrm{bxy}}} = C_{\omega^{\mathrm{bxy}}} \sum_{i\in(x,y)} {\omega^{\mathrm{b}_i}}^2. \quad (6)$$

The orientation penalty term penalizes all those robot base orientations $\boldsymbol{\phi}^{\mathrm{b}}$ that do not make the gravity vector $\boldsymbol{g} = \langle 0, 0, -9.81 \rangle$ perpendicular to the x–y plane of the robot base frame by keeping the x and y components of the projected gravity vector $\boldsymbol{\phi}^{\mathrm{b}}_{\|g} = \frac{\boldsymbol{\phi}^{\mathrm{b}} \cdot \boldsymbol{g}}{\|\boldsymbol{g}\|^2} \boldsymbol{g} = \langle p_{\mathrm{g}}^{\mathrm{b_x}}, p_{\mathrm{g}}^{\mathrm{b_y}}, p_{\mathrm{g}}^{\mathrm{b_z}} \rangle$ as close to zero as possible

$$p_{p_{\mathrm{g}}^{\mathrm{bxy}}} = C_{p_{\mathrm{g}}^{\mathrm{bxy}}} \sum_{i\in(x,y)} {p_{\mathrm{g}}^{\mathrm{b}_i}}^2. \quad (7)$$

The robot base height penalty term keeps the robot base height, with regard to the terrain beneath it, $p^{\mathrm{b_z}}$ close to a predefined walking height h (0.52 m) to avoid the robot walking with a low base height (this penalty term is particularly helpful when training blind locomotion policies on rough terrains, as is shown in the results):

$$p_{p^{\mathrm{b_z}}} = C_{p^{\mathrm{b_z}}} (p^{\mathrm{b_z}} - h)^2. \quad (8)$$

The joint torques penalty term aims at minimizing the robot actuated joint torques for energy-efficiency reasons:

$$p_{\tau} = C_{\tau} \sum_{i=1}^{n} \tau_i^2. \quad (9)$$

The joint accelerations penalty term aims at minimizing the robot actuated joint accelerations to avoid abrupt motions of the robot joints:

$$p_{\ddot{q}} = C_{\ddot{q}} \sum_{i=1}^{n} (\ddot{q}_i^{\mathrm{t}-1} - \ddot{q}_i^{\mathrm{t}})^2. \quad (10)$$

The action rate penalty term smooths the policy output with regard to time steps, providing a certain continuity in the desired joint references regardless of whether they are joint positions or joint velocities:

$$p_{\dot{a}} = C_{\dot{a}} \sum_{i=1}^{n} (a_i^{\mathrm{t}-1} - a_i^{\mathrm{t}})^2. \quad (11)$$

The hip motion penalty term is used for aesthetic reasons, and it keeps the Hip Abduction–Adduction joint positions close to the default values, which makes the robot walk in a natural-looking manner:

$$p_{\dot{q}_{\mathrm{H}}} = C_{\dot{q}_{\mathrm{H}}} \sum_{\substack{i \in (L,R) \\ j \in (F,H)}} |q_{ij\_\mathrm{HAA}} - q_{ij\_\mathrm{HAA}}^{\mathrm{def}}|, \quad (12)$$

where L, R, F, and H are the left, right, front, and hind sides of the robot, respectively, and $q_{ij\_\mathrm{HAA}}^{\mathrm{def}}$ are the default joint positions of the Hip Abduction–Adduction joints, which are 0.03 radians for the left-hand side joints and -0.03 radians for the right-hand side joints.

In Equations (5)–(12), C_* are opportunely tuned constants, and n is the number of robot actuated joints. These reward and penalty terms do not impose any kind of specific walking pattern, gait, or behavior, neither static nor dynamic, since none of them uses any kind of information about feet contact timing. These terms only aim to impose a specific robot base motion, since they only involve robot base velocity, orientation, and height values, as well as joint velocities, accelerations, and torques. The dynamic locomotion gait is a self-emerging behavior during training. According to these terms, it is more convenient to employ a faster dynamic gait rather than a slower static one to track the required commanded base twists to obtain a higher reward.

The robots are reset and their rewards are penalized if their base or more than one knee gets in touch with the ground. Robots can also be reset if a timeout occurs when reaching the maximum episode length. Timeouts are correctly handled, as in [14], using the partial-episode bootstrapping method [40]. The values of the hyper-parameters of the policy optimization algorithm (PPO), used for the training of the locomotion policies, are the environment default ones as in [35].

The policies were all trained in simulation using Isaac Gym [13] in the Omniverse Isaac Sim [34] simulator, running thousands of robots in parallel, exploiting the power of GPU acceleration. The hardware setup for training the control policies employed an NVIDIA Tesla T4 GPU with 16 GB of VRAM.

4. Results

In Figure 7, the reward trends of the different training sessions are shown. The spikes in the trends are due to training restarts from checkpoints. The training was stopped when the reward did not significantly increase for more than a thousand iterations. The number of simultaneously simulated robots during the training process is 2048 for all the trained policies. A policy update iteration consists of 48 policy steps for each of these environments.

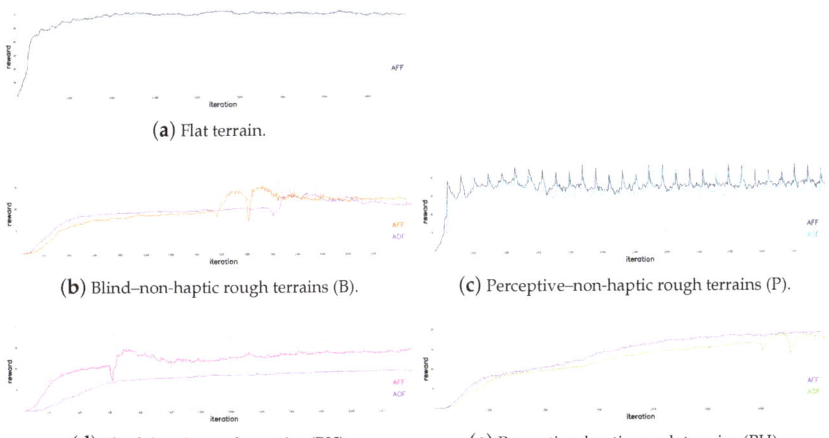

(**a**) Flat terrain.

(**b**) Blind–non-haptic rough terrains (B).

(**c**) Perceptive–non-haptic rough terrains (P).

(**d**) Blind–haptic rough terrains (BH).

(**e**) Perceptive–haptic rough terrains (PH).

Figure 7. Reward trends of the different training sessions. AFF stands for Adaptive Flat Feet (Figure 3b), while AOF stands for Adaptive Open Feet (Figure 3c).

The performance indices evaluated are the following: the linear velocity tracking error in the forward and lateral directions; the yaw angular velocity tracking error around the vertical axis of the robot base frame; the desired joint positions rate, which is the difference between the previous policy output and the current one, to evaluate continuity; the actuated joint velocities rate, which is the difference between the previous actuated joint velocities and the current ones, to evaluate abrupt motions; actuated joint torques, to evaluate energy efficiency; and robot base height, to evaluate natural-looking walking style.

After the policies were trained, the performance indices were evaluated by simulating 1024 robots in parallel for a simulation time of 60 s for the flat terrain locomotion task and 17.5 s for the rough terrain policies. This is just for memory requirements and computational time reasons for computing these performance indices. Both the mean and standard deviation of the performance indices were computed over all these policy steps, and they are shown in the charts in terms of error bars.

Figure 8 shows a performance evaluation and a comparison of the flat terrain locomotion policies. The performance of the baseline policy from [35], employed on the robot equipped with its own default ball feet, is compared with the performance of the same policy employed on the robot equipped with the Adaptive Flat Feet (AFF, Figure 3b). Then the performance of a third policy, opportunely trained using the robot equipped with the Adaptive Flat Feet, is also illustrated. A detailed description of the different policies tested on this task is given in Table 2.

Table 2. Description of the different robot setups tested for the flat terrain locomotion task.

Robot Setup	Description
ANY	**ANY**mal-C with its default ball feet
AFF-BFP	ANYmal-C with **A**daptive **F**lat **F**eet controlled by the **B**all **F**eet **P**olicy
AFF	ANYmal-C with **A**daptive **F**lat **F**eet

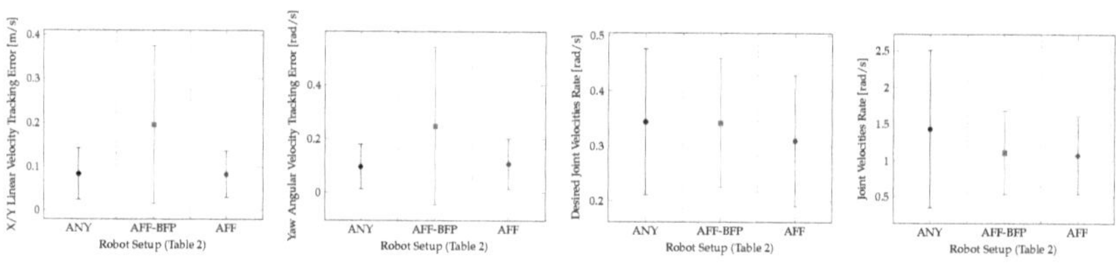

Figure 8. Performance of the different robot setups tested for the flat terrain locomotion task.

In Figure 9, a performance evaluation of the rough terrain locomotion policies and comparison with ball-feet baseline policies from [35] is presented. Furthermore, in this case, the baseline policy (perceptive–non-haptic with ball feet P-ANY) from [35], trained on the robot equipped with its own default ball feet, was tested on the robot equipped with both the approximations of the SoftFoot-Q. A detailed description of the different policies tested on this task is given in Table 3.

Table 3. Description of the different robot setups tested for the rough terrain locomotion task.

Robot Setup	Description
B-ANY	**B**lind–non-haptic **ANY**mal-C with its default ball feet
B-AFF	**B**lind–non-haptic ANYmal-C with **A**daptive **F**lat **F**eet
B-AOF	**B**lind–non-haptic ANYmal-C with **A**daptive **O**pen **F**eet
P-ANY	**P**erceptive–non-haptic **ANY**mal-C with its default ball feet
P-AFF-BFP	**P**erceptive–non-haptic ANYmal-C with **A**daptive **F**lat **F**eet controlled by the **B**all **F**eet **P**olicy
P-AOF-BFP	**P**erceptive–non-haptic ANYmal-C with **A**daptive **O**pen **F**eet controlled by the **B**all **F**eet **P**olicy
P-AFF	**P**erceptive–non-haptic ANYmal-C with **A**daptive **F**lat **F**eet
P-AOF	**P**erceptive–non-haptic ANYmal-C with **A**daptive **O**pen **F**eet
BH-AFF	**B**lind–**H**aptic ANYmal-C with **A**daptive **F**lat **F**eet
BH-AOF	**B**lind–**H**aptic ANYmal-C with **A**daptive **O**pen **F**eet
PH-AFF	**P**erceptive–**H**aptic ANYmal-C with **A**daptive **F**lat **F**eet
PH-AOF	**P**erceptive–**H**aptic ANYmal-C with **A**daptive **O**pen **F**eet

For the blind locomotion tasks, we noticed a common pattern: the robot learned to walk at a lower height of the base. As mentioned in Section 3, adding the base height penalty term partially helps to solve this problem at the expense of a worse base twist command tracking performance.

Since the policies were trained taking into account dynamics randomization, as discussed in Section 3, Figure 10 shows an example of the disturbance response to random pushes applied to the robot base for the perceptive–non-haptic rough terrain policy running on the robot equipped with the Adaptive Flat Feet (AFF).

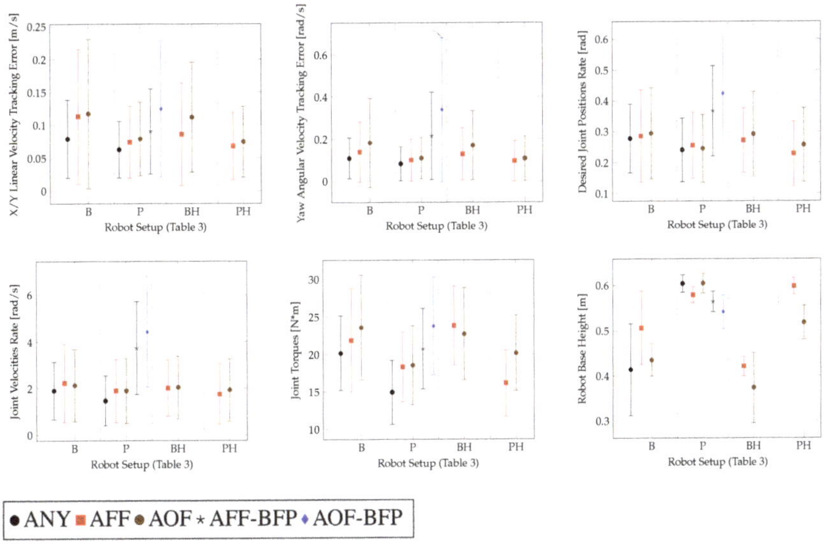

Figure 9. Performance of the different robot setups tested for the rough terrain locomotion task.

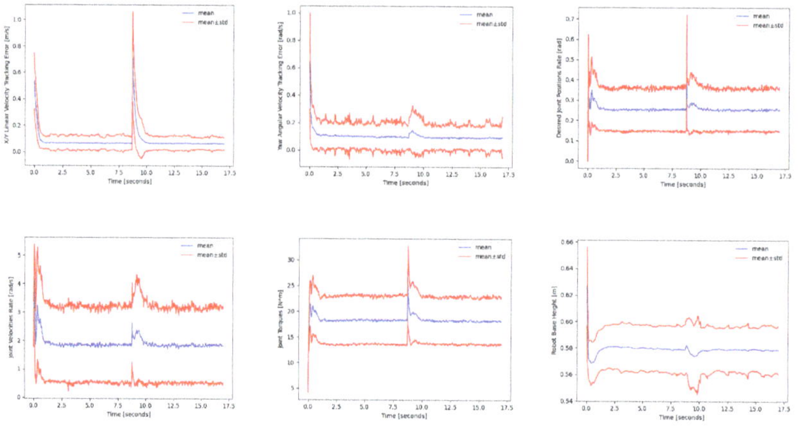

Figure 10. Performance evaluation when random pushes are applied to the robot base. The disturbance is applied at 8.75 s. The mean and the standard deviation were computed by simulating 1024 robots in parallel, with each one following a random commanded base twist reference.

Figures 11–14 show two motion captures of two different trained policies traversing two types of rough terrains among the considered ones and their corresponding feet contact schedules. Please refer also to the video attachment to visualize the full experiment. The feet contact schedules in Figures 12 and 14 show that the learned locomotion gait follows a dynamic gait pattern, namely, the trotting gait.

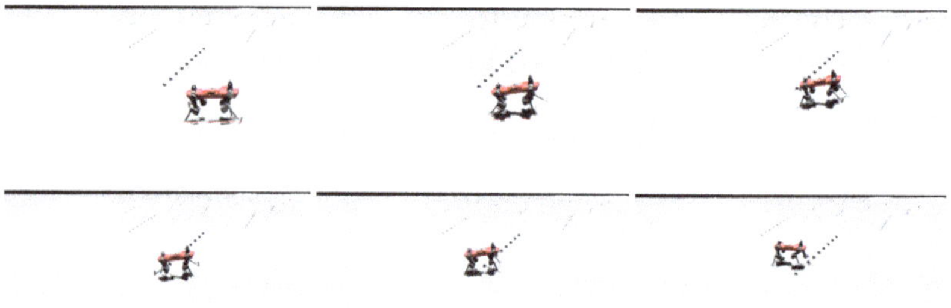

Figure 11. Photo sequence of the P-AFF policy on stairs.

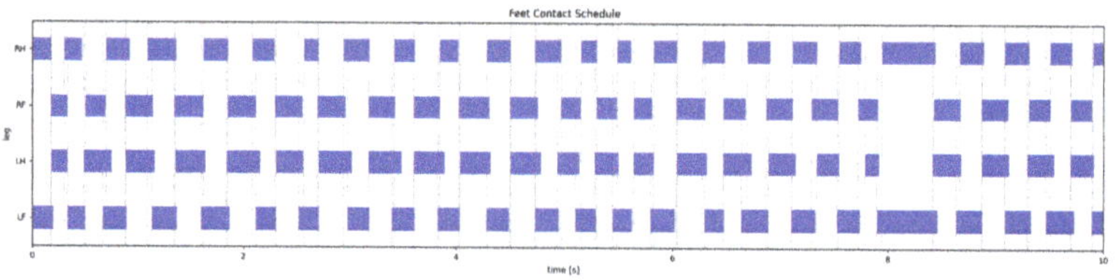

Figure 12. Feet contact schedule of the P-AFF policy on stairs.

Figure 13. Photo sequence of the PH-AOF policy on a rough slope.

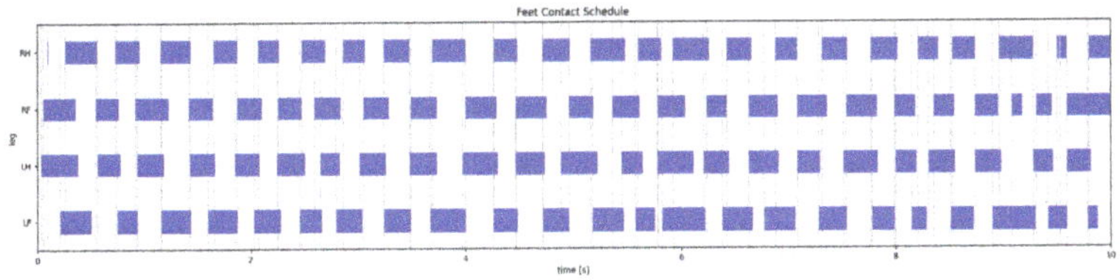

Figure 14. Feet contact schedule of the PH-AOF policy on a rough slope.

5. Discussion

For the flat terrain locomotion task, only the Adaptive Flat Foot (AFF) approximation was considered, because on this type of terrain, the adaptability features of the SoftFoot-Q would not be exploited. In this case, we obtain similar results with regard to the robot equipped with ball feet. The performance of the flat terrain and perceptive–non-haptic rough terrain (P) baseline policies from [35], trained using the robot equipped with its own default ball feet (ANY), decreases when running on the robot equipped with either the Adaptive Flat Feet (AFF) or the Adaptive Open Feet (AOF) approximation. This shows that the same policy trained using ball feet cannot be employed with the adaptive feet. This proves that appropriate training is required to reach the same level of performance as in the ball feet case (ANY), as shown in Figures 8 and 9. We can conclude that with adaptive feet, we reach a similar performance with regard to the flat terrain and perceptive–non-haptic rough terrain (P) baseline policies employing ball feet (ANY) [35].

In the cases of blind rough terrain locomotion (B, BH), the robot is led to walk with a lower height of the base, but it is still able to track the base twist reference command in a comparable manner to the perceptive-non-haptic (P) setup. This phenomenon is more evident in the case of ball feet (B-ANY). In Figure 9, we can see that with adaptive feet (B-AFF, B-AOF), the twist command tracking performance is similar to the ball feet case (B-ANY), but the robot base height is slightly higher and more stable given its lower variation. Therefore, in the blind settings, the walk is more resilient with adaptive feet [10].

We can also see that the Adaptive Open Foot (AOF) approximation, in the cases of blind locomotion on rough terrains, provides, in general, worse performance with regard to the Adaptive Flat Foot (AFF) approximation.

Finally, there are orientation problems with foot placement. On rough terrains, we experienced a lot of cases in which the feet assumed a configuration that would cause a break of the foot on the real robot, especially with blind locomotion, like walking while making contact with the ground either with the lateral sides of the feet or with the upper side of the front and back arch links. This problem can be seen in Figure 15.

Figure 15. Foot orientation problem.

6. Conclusions & Future Works

In this work, we presented blind, perceptive, and haptic end-to-end learning-based policies for dynamic locomotion control of quadrupedal robots equipped with the adaptive soft feet in [10]. We presented the training setup and approach, with a detailed evaluation of the performance obtained and a comparison with existing baselines. We investigated the benefits and the faults.

Some of the future work that we aim to put our efforts into are experiments on the real robot, to better assess the goodness of the adaptive feet model approximations and validate them, and performing training using the foothold cost map instead of the elevation map. This second goal could be achieved through our previous work [26], in which we designed and developed a CNN-based foothold selection algorithm for static gaits that works with adaptive feet. Given the dynamic locomotion problem complexity, the selected footholds cannot be taken into account separately by planning static gaits, but the cost maps related to each foot need to be processed at the same time to guarantee convergence and good performance.

Moreover, model-based methods could also be used together with data-driven methods [41] to further improve the controller reliability and overall performance.

Author Contributions: Conceptualization, A.S., F.A. and M.G.; Methodology, A.S., F.A. and M.G.; Software, A.S.; Validation, A.S.; Formal analysis, A.S.; Investigation, A.S., F.A. and M.G.; Data curation, A.S.; Writing—original draft, A.S.; Writing—review and editing, F.A. and M.G.; Visualization, A.S.; Supervision, F.A. and M.G.; Project administration, M.G.; Funding acquisition, M.G. All authors have read and agreed to the published version of the manuscript..

Funding: This research is partially supported by the European Union's Horizon 2020 Research and Innovation Programme under Grant Agreement No. 101016970 (Natural Intelligence), in part by the Ministry of University and Research (MUR) as part of the PON 2014-2020 "Research and Innovation" resources – Green/Innovation Action - DM MUR 1061/2021 and DM MUR 1062/2021, and in part by the Italian Ministry of Education and Research (MIUR) in the framework of the FoReLab project (Departments of Excellence).

Data Availability Statement: The data presented in this study are available on request from the corresponding author.

Conflicts of Interest: The authors declare no conflicts of interest.

References

1. Kim, S.; Wensing, P.M. Design of dynamic legged robots. *Found. Trends Robot.* **2017**, *5*, 117–190. [CrossRef]
2. Fankhauser, P.; Hutter, M. Anymal: A unique quadruped robot conquering harsh environments. *Res. Features* **2018** , *126*, 54–57. [CrossRef]
3. Wong, C.; Yang, E.; Yan, X.T.; Gu, D. Autonomous robots for harsh environments: A holistic overview of current solutions and ongoing challenges. *Syst. Sci. Control. Eng.* **2018**, *6*, 213–219. [CrossRef]
4. Fahmi, S.; Mastalli, C.; Focchi, M.; Semini, C. Passive whole-body control for quadruped robots: Experimental validation over challenging terrain. *IEEE Robot. Autom. Lett.* **2019**, *4*, 2553–2560. [CrossRef]
5. Liu, J.; Tan, M.; Zhao, X. Legged robots—An overview. *Trans. Inst. Meas. Control* **2007**, *29*, 185–202. [CrossRef]

6. Sun, J.; Wang, Z.; Zhang, M.; Zhang, S.; Qian, Z.; Chu, J. A novel metamorphic foot mechanism with toe joints based on spring-loaded linkages. *IEEE Robot. Autom. Lett.* **2022**, *8*, 97–104. [CrossRef]
7. Chatterjee, A.; Mo, A.; Kiss, B.; Gönen, E.C.; Badri-Spröwitz, A. Multi-segmented adaptive feet for versatile legged locomotion in natural terrain. In Proceedings of the 2023 IEEE International Conference on Robotics and Automation (ICRA), London, UK, 29 May–2 June 2023; pp. 1162–1169.
8. Ranjan, A.; Angelini, F.; Nanayakkara, T.; Garabini, M. Design Guidelines for Bioinspired Adaptive Foot for Stable Interaction With the Environment. *IEEE/ASME Trans. Mechatron.* **2024**, *29*, 843–855. [CrossRef]
9. Yao, C.; Shi, G.; Xu, P.; Lyu, S.; Qiang, Z.; Zhu, Z.; Ding, L.; Jia, Z. STAF: Interaction-based design and evaluation of sensorized terrain-adaptive foot for legged robot traversing on soft slopes. *IEEE/ASME Trans. Mechatron.* **2024** , 1–12.
10. Catalano, M.G.; Pollayil, M.J.; Grioli, G.; Valsecchi, G.; Kolvenbach, H.; Hutter, M.; Bicchi, A.; Garabini, M. Adaptive Feet for Quadrupedal Walkers. *IEEE Trans. Robot.* **2022**, *38*, 302–316. [CrossRef]
11. Aoi, S.; Tanaka, T.; Fujiki, S.; Funato, T.; Senda, K.; Tsuchiya, K. Advantage of straight walk instability in turning maneuver of multilegged locomotion: A robotics approach. *Sci. Rep.* **2016**, *6*, 30199. [CrossRef]
12. Holmes, P.; Full, R.J.; Koditschek, D.; Guckenheimer, J. The Dynamics of Legged Locomotion: Models, Analyses, and Challenges. *SIAM Rev.* **2006**, *48*, 207–304. [CrossRef]
13. Makoviychuk, V.; Wawrzyniak, L.; Guo, Y.; Lu, M.; Storey, K.; Macklin, M.; Hoeller, D.; Rudin, N.; Allshire, A.; Handa, A.; et al. Isaac Gym: High Performance GPU Based Physics Simulation For Robot Learning. In Proceedings of the Thirty-fifth Conference on Neural Information Processing Systems Datasets and Benchmarks Track (Round 2), Virtual-only Conference, 6–14 December 2021.
14. Rudin, N.; Hoeller, D.; Reist, P.; Hutter, M. Learning to Walk in Minutes Using Massively Parallel Deep Reinforcement Learning. In Proceedings of the 5th Conference on Robot Learning, PMLR, London, UK, 8–11 November 2022. In Proceedings of Machine Learning Research; Faust, A., Hsu, D., Neumann, G., Eds.; PMLR: London, UK, 2022; Volume 164, pp. 91–100.
15. Iscen, A.; Caluwaerts, K.; Tan, J.; Zhang, T.; Coumans, E.; Sindhwani, V.; Vanhoucke, V. Policies modulating trajectory generators. In Proceedings of the 2nd Conference on Robot Learning, PMLR , 2018; Zürich, Switzerland, 29–31 October, pp. 916–926.
16. Lee, J.; Hwangbo, J.; Wellhausen, L.; Koltun, V.; Hutter, M. Learning quadrupedal locomotion over challenging terrain. *Sci. Robot.* **2020**, *5*, eabc5986. [CrossRef] [PubMed]
17. Hwangbo, J.; Lee, J.; Dosovitskiy, A.; Bellicoso, D.; Tsounis, V.; Koltun, V.; Hutter, M. Learning agile and dynamic motor skills for legged robots. *Sci. Robot.* **2019**, *4*, eaau5872. [CrossRef] [PubMed]
18. Bai, S.; Kolter, J.Z.; Koltun, V. An empirical evaluation of generic convolutional and recurrent networks for sequence modeling. *arXiv* **2018**, arXiv:1803.01271.
19. Miki, T.; Lee, J.; Hwangbo, J.; Wellhausen, L.; Koltun, V.; Hutter, M. Learning robust perceptive locomotion for quadrupedal robots in the wild. *Sci. Robot.* **2022**, *7*, eabk2822. [CrossRef]
20. Arena, P.; Li Noce, A.; Patanè, L. Stability and Safety Learning Methods for Legged Robots. *Robotics* **2024**, *13*, 17. [CrossRef]
21. Rudin, N.; Hoeller, D.; Bjelonic, M.; Hutter, M. Advanced Skills by Learning Locomotion and Local Navigation End-to-End. In Proceedings of the 2022 IEEE/RSJ International Conference on Intelligent Robots and Systems (IROS), Kyoto, Japan, 23–27 October 2022; pp. 2497–2503. [CrossRef]
22. Yang, R.; Zhang, M.; Hansen, N.; Xu, H.; Wang, X. Learning Vision-Guided Quadrupedal Locomotion End-to-End with Cross-Modal Transformers. *arXiv* **2022**, arXiv:2107.03996 Available online: http://arxiv.org/abs/2107.03996 (accessed on 23 June 2024).
23. Wu, J.; Xin, G.; Qi, C.; Xue, Y. Learning Robust and Agile Legged Locomotion Using Adversarial Motion Priors. *IEEE Robot. Autom. Lett* **2023**, *8*, 4975–4982. [CrossRef]
24. Hoeller, D.; Rudin, N.; Sako, D.; Hutter, M. ANYmal parkour: Learning agile navigation for quadrupedal robots. *Sci. Robot.* **2024**, *9*, eadi7566. [CrossRef]
25. Margolis, G.B.; Yang, G.; Paigwar, K.; Chen, T.; Agrawal, P. Rapid locomotion via reinforcement learning. *Int. J. Robot. Res.* **2024**, *43*, 572–587. [CrossRef]
26. Bednarek, J.; Maalouf, N.; Pollayil, M.J.; Garabini, M.; Catalano, M.G.; Grioli, G.; Belter, D. CNN-based Foothold Selection for Mechanically Adaptive Soft Foot. In Proceedings of the 2020 IEEE/RSJ International Conference on Intelligent Robots and Systems (IROS), Las Vegas, NV, USA, 24 October 2020–24 January 2021; pp. 10225–10232. [CrossRef]
27. Belter, D.; Bednarek, J.; Lin, H.C.; Xin, G.; Mistry, M. Single-shot Foothold Selection and Constraint Evaluation for Quadruped Locomotion. In Proceedings of the 2019 International Conference on Robotics and Automation (ICRA), Montreal, QC, Canada, 20–24 May 2019; pp. 7441–7447. [CrossRef]
28. Belter, D.; Skrzypczyński, P. Rough terrain mapping and classification for foothold selection in a walking robot. *J. Field Robot.* **2011**, *28*, 497–528. [CrossRef]
29. Romera, E.; Álvarez, J.M.; Bergasa, L.M.; Arroyo, R. ERFNet: Efficient Residual Factorized ConvNet for Real-Time Semantic Segmentation. *IEEE Trans. Intell. Transp. Syst.* **2018**, *19*, 263–272. [CrossRef]
30. Magaña, O.A.V.; Barasuol, V.; Camurri, M.; Franceschi, L.; Focchi, M.; Pontil, M.; Caldwell, D.G.; Semini, C. Fast and Continuous Foothold Adaptation for Dynamic Locomotion Through CNNs. *IEEE Robot. Autom. Lett.* **2019**, *4*, 2140–2147. [CrossRef]
31. Xin, G.; Xin, S.; Cebe, O.; Pollayil, M.J.; Angelini, F.; Garabini, M.; Vijayakumar, S.; Mistry, M. Robust Footstep Planning and LQR Control for Dynamic Quadrupedal Locomotion. *IEEE Robot. Autom. Lett.* **2021**, *6*, 4488–4495. [CrossRef]

32. Schulman, J.; Wolski, F.; Dhariwal, P.; Radford, A.; Klimov, O. Proximal Policy Optimization Algorithms. *arXiv* **2017**, arXiv:1707.06347. Availble online: http://arxiv.org/abs/1707.06347 (accessed on 23 June 2024).
33. Makoviichuk, D.; Makoviychuk, V. rl-games: A High-performance Framework for Reinforcement Learning. 2021. Available online: https://github.com/Denys88/rl_games (accessed on 23 June 2024).
34. NVIDIA. Isaac Sim—Robotics Simulation and Synthetic Data Generation. 2022. Available online: https://developer.nvidia.com/isaac-sim (accessed on 23 June 2024).
35. NVIDIA-Omniverse. Omniverse Isaac Gym Reinforcement Learning Environments for Isaac Sim. 2022. Available online: https://github.com/NVIDIA-Omniverse/OmniIsaacGymEnvs (accessed on 23 June 2024).
36. Hutter, M.; Gehring, C.; Jud, D.; Lauber, A.; Bellicoso, C.D.; Tsounis, V.; Hwangbo, J.; Bodie, K.; Fankhauser, P.; Bloesch, M.; et al. ANYmal - a highly mobile and dynamic quadrupedal robot. In Proceedings of the 2016 IEEE/RSJ International Conference on Intelligent Robots and Systems (IROS), Daejeon, Republic of Korea, 9–14 October 2016; pp. 38–44. [CrossRef]
37. Bengio, Y.; Louradour, J.; Collobert, R.; Weston, J. Curriculum Learning. In Proceedings of the 26th Annual International Conference on Machine Learning, ICML '09, New York, NY, USA , 2009; 14-18 June, pp. 41–48. [CrossRef]
38. Wang, R.; Lehman, J.; Clune, J.; Stanley, K.O. Paired Open-Ended Trailblazer (POET): Endlessly Generating Increasingly Complex and Diverse Learning Environments and Their Solutions. *arXiv* **2019**, arXiv:1901.01753. Availble online: http://arxiv.org/abs/1901.01753 (accessed on 23 June 2024).
39. Peng, X.B.; Andrychowicz, M.; Zaremba, W.; Abbeel, P. Sim-to-Real Transfer of Robotic Control with Dynamics Randomization. In Proceedings of the 2018 IEEE International Conference on Robotics and Automation (ICRA), Brisbane, QLD, Australia, 21–25 May 2018; pp. 3803–3810. [CrossRef]
40. Pardo, F.; Tavakoli, A.; Levdik, V.; Kormushev, P. Time Limits in Reinforcement Learning. In Proceedings of the 35th International Conference on Machine Learning, PMLR, Stockholm, Sweden, 10–15 July 2018; Proceedings of Machine Learning Research; Dy, J., Krause, A., Eds.; PMLR: Stockholm, Sweden, 2018; Voume 80, pp. 4045–4054.
41. Arena, P.; Patanè, L.; Taffara, S. A Data-Driven Model Predictive Control for Quadruped Robot Steering on Slippery Surfaces. *Robotics* **2023**, *12*, 67. [CrossRef]

robotics MDPI

Article

Boosting Deep Reinforcement Learning with Semantic Knowledge for Robotic Manipulators

Lucía Güitta-López [1,*,†], Vincenzo Suriani [2,†], Jaime Boal [1], Álvaro J. López-López [1] and Daniele Nardi [3]

[1] Institute for Research in Technology (IIT), ICAI School of Engineering, Comillas Pontifical University, Rey Francisco, 4, 28008 Madrid, Spain; jaime.boal@iit.comillas.edu (J.B.); alvaro.lopez@iit.comillas.edu (Á.J.L.-L.)
[2] Department of Engineering, University of Basilicata, Via dell'Ateneo Lucano, 10, 85100 Potenza, Italy; vincenzo.suriani@unibas.it
[3] Department of Computer, Control and Management Engineering, Sapienza University of Rome, Via Ariosto, 25, 00185 Rome, Italy; nardi@diag.uniroma1.it
* Correspondence: lucia.guitta@iit.comillas.edu
† These authors contributed equally to this work.

Abstract

Deep Reinforcement Learning (DRL) is a powerful framework for solving complex sequential decision-making problems, particularly in robotic control. However, its practical deployment is often hindered by the substantial amount of experience required for learning, which results in high computational and time costs. In this work, we propose a novel integration of DRL with semantic knowledge in the form of Knowledge Graph Embeddings (KGEs), aiming to enhance learning efficiency by providing contextual information to the agent. Our architecture combines KGEs with visual observations, enabling the agent to exploit environmental knowledge during training. Experimental validation with robotic manipulators in environments featuring both fixed and randomized target attributes demonstrates that our method achieves up to 60% reduction in learning time and improves task accuracy by approximately 15 percentage points, without increasing training time or computational complexity. These results highlight the potential of semantic knowledge to reduce sample complexity and improve the effectiveness of DRL in robotic applications.

Keywords: deep reinforcement learning; semantic knowledge; robotics; sample efficiency

1. Introduction

Deep Reinforcement Learning (DRL) [1,2] is now consolidating as one of the most promising solutions for solving complex sequential decision problems where an agent interacts with its environment. While previous reinforcement learning (RL) [3] is limited by its memory and computational capabilities, DRL overcomes these limitations by leveraging the potential of deep learning for function approximation and representation learning in high-dimensional state spaces [4,5].

The application of DRL to tasks performed by robotic manipulators has broadened the scope from mostly repetitive jobs in fixed scenarios to tasks in changing environments with potentially unknown dynamics [6–9]. One of the major disadvantages of DRL is the extensive amount of experience, i.e., interactions between the agent and the environment, required for learning. It should also be noted that the greater the complexity of the environment, the more experience is needed. Virtual environments can be used to reduce the time and economic cost of overruns. By modeling assets in virtual scenarios, the *sample efficiency* problem is mitigated, as experience is acquired more quickly in simulators. However, the issue of the sheer amount of experience required remains.

As Figure 1 presents, we propose integrating contextual information about the environment within the agent's learning architecture to address the sample efficiency issue and enhance the agent's performance. Our main motivation is that any learning process, especially in DRL setups, should improve if the agent has knowledge about the environment that complements its observations. As already shown in [10], providing some basic semantic structure to the agent can speed up its learning. Ultimately, we argue that this approach introduces *common sense* as semantic knowledge without additional complexity cost, with only a slight increase in the computational time. The environment's semantic information is obtained from a pre-built knowledge graph, where we extract the embeddings. After the representation learning layers, these embeddings are concatenated as inputs in the agent's architecture. To sum up, our main contributions and findings are the following:

- An original DRL agent architecture that successfully integrates Knowledge Graph Embeddings (KGEs) during the learning process. We modify the baseline layers to allow for concatenating the KGEs with the hidden activations from the "visual layers".
- A methodology to quantify the improvement of the DRL agents' performance, in terms of time and accuracy, through the use of KGEs with respect to a baseline agent with no semantic information of the environment and an agent with only partial information.
- An analysis of the possible improvement of the agents' exploration and exploitation capabilities by examining the distribution of each joint orientation during the evaluation episodes.
- A quantitative and qualitative evaluation of the embedding's influence in different environments and with various robot manipulators.

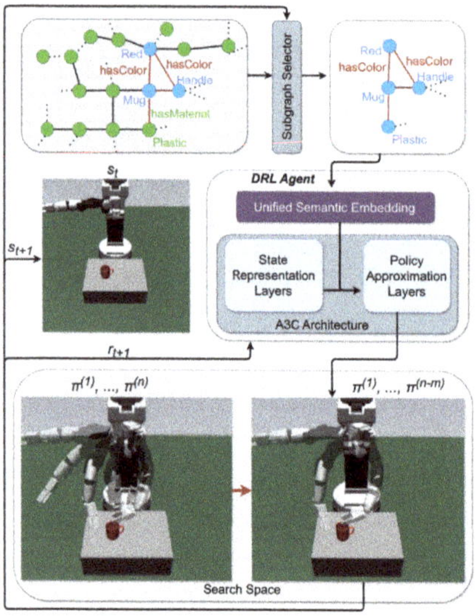

Figure 1. Proposed framework diagram. A subgraph selector is used to extract the important entities and relationships based on the environment from a complete graph. This subgraph is embedded and concatenated in the layer prior to the policy approximation blocks of the DRL agent architecture. The result is an improvement in learning times and accuracy and a reduction in the required exploration.

The paper is structured as follows: Section 2 presents the related work and Section 3 defines the experiments conducted, the setup employed, and the MDP designed. In Section 4, we describe the methodology followed for the knowledge graph and embedding obtention, as well as the Deep Reinforcement Learning framework; that is, the algorithm used and our architecture, besides the corresponding training and post-training evaluation procedures. Section 5 gathers the results discussed in Section 6. Section 7 focuses on describing shortcomings and open issues related to our approach, and finally, Section 8 sums up the most relevant insights obtained with this research.

2. Related Work

To leverage the capability of robots to interact with the physical world, robot learning has emerged as a major research area [11]. Progress in deep learning has contributed significantly to advancements in robot learning, particularly when state representation includes images [12]. A variety of algorithms have been developed for robot learning, with reinforcement learning and imitation learning remaining the dominant approaches [13,14]. To facilitate the comparison of different reinforcement learning algorithms, numerous benchmarks have been introduced. For instance, some benchmarks focus on specific tasks such as door opening, furniture assembly, and in-hand dexterous manipulation [15]. Others, like those in [16,17], offer diverse environments but lack long-horizon tasks and lack of background knowledge. Background knowledge has been successfully deployed when dealing with natural language action space, to reduce the search space [18,19], but not extensively applied in the field of robotics.

Recently, common sense has been integrated into the RL pipelines to enrich the representation of the environments. With the advent of the LLMs, a taxonomy of the combination of the two fields of study has been proposed in [20], where three classes are defined based on the way that the two model types interact with each other. An LLM can be used to supplement the training of an RL model that performs a general task that is not inherently related to natural language, as *LLM4RL* states.

Here, the semantic knowledge of Large Language Models (LLMs) is leveraged to enhance the performance of reinforcement learning (RL) agents [21]. This enhancement can occur in two primary ways: by grounding the agent's environment to improve its performance [22–24], or by improving the training process of the RL agent. Training an RL agent is often computationally intensive and requires significant resources and large amounts of data. Moreover, RL training can suffer from inefficient sampling, particularly for complex, long-term tasks. Consequently, several LLM4RL frameworks aim to enhance training efficiency, ensuring the successful execution of target tasks during testing. They achieve this by facilitating exploration [25,26], enabling policy transfer of trained models [27], and implementing effective planning to reduce data requirements.

The role of a large-scale model in helping an RL agent is discussed in [28], where the authors investigate how to combine these complementary abilities in a single system consisting of three parts: a Planner, an Actor, and a Reporter.

In the learning process of a grasping task, it has been demonstrated how generating natural human grasps needs to consider not only the object geometry, but also semantic information. In fact, in [29], a semantic-based grasp generation method, termed SemGrasp, generates a static human grasp pose by incorporating semantic information into the grasp representation.

If LLMs offer an invaluable opportunity to provide semantic information, it is also true that they can be demanding in many applications. To this end, we adopted a lighter architecture, using graphs. Semantic knowledge can successfully be stored in graph representations and manipulated using Knowledge Graph Embedding techniques and pretrained models. Some pioneering work has employed research-rich reinforcement learning (RL) techniques to address graph mining tasks. In [30], a unified graph reinforcement learning (GRL) formulation is proposed. However, most existing formulations struggle to effectively integrate groups of semantically related nodes within the RL paradigm. On the other hand, far from the RL field, a semantic representation framework based on a knowledge graph has been demonstrated to be suitable as a method to extract manipulation knowledge from multiple sources of information [31] and guide the agent to generate semantic representations of entities and relations in the knowledge base.

We propose to leverage the semantic knowledge of a lightweight graph representation to improve sample efficiency and agent performance in terms of time and accuracy. Our architecture provides additional contextual information about the environment, which complements the agent's observations, leading to faster learning and higher accuracy, compared to baseline models without knowledge graphs.

3. Environment Setup

3.1. Experiment Description

The task addressed involves using a robotic arm to approach three targets (e.g., a mug, a bottle, and a cereal box) that move randomly in the workspace area between episodes. The robot's end-effector should reach the grasping point within a distance of 5 cm during training and 10 cm during the post-training evaluation, with a specific orientation. The training orientation threshold is $\pm 15^{\circ}$, whereas in the post-training evaluation, it is $\pm 20^{\circ}$. The placement of the reachable point and the gripper approach orientation depend on the target. The environment observation is a $64 \times 64 \times 3$ RGB image without additional proprioception information. We performed the experiments in simulation to enable a thorough analysis of the proposed approach and to allow for total control of all the external drivers that might affect the agent's performance. In this way, we ensure total control of all the external drivers that might decrease the agents' performance.

3.2. Setup Description

Two robots have been used in this research: on the one hand, the TIAGo from PAL Robotics [32], a 7-DoF mobile manipulator equipped with a two-finger gripper and mainly designed for human-robot interaction; on the other hand, the IRB120 [33] from ABB, a 6-DoF industrial robotic arm also with a two-finger gripper end-effector. The virtualization of both assets has been carried out through the MuJoCo [34] platform. The arena is the same for both robots. Due to the morphology of the TIAGo arm and to facilitate the reach, it was necessary to add a table that allowed us to place the objects at a greater height in the robot's workspace. Figure 2 displays the virtual environment for the TIAGo and IRB120 models.

The targets are represented by three graspable objects: a mug, a bottle, and a cereal box. Figure 2 shows the reachable spots for the three targets. We present the grasping point as a black sphere for better understanding of the reader, but in the observation received by the agent, it is not shown. Moreover, Figure 2 illustrates the orientations of the grasping point axes that the grippers must align with, where the x coordinate is red, the y coordinate is green, and the z coordinate is blue. In the case of the mug, the grasping point is placed on its handle within an inclined orientation with respect to the base plane. For the bottle, we placed the grasping point on the stopper perpendicular to the base plane. Lastly, in the cereal box, the grasping point is set on the right-hand-side edge, parallel to the base plane, above half its height. For the experiments without domain randomization (DR) applied to the targets' color, the mug is red (RGB code (1.0, 0.0, 0.0)), the bottle is yellow (RGB code (1.0, 1.0, 0.0)), and the cereal box is brown (RGB code (0.55, 0.27, 0.07)). When we experiment with DR applied to the targets' color, the mug can be red or blue (RGB code (0.0, 0.0, 1.0)), the bottle yellow or purple (RGB code (0.4, 0.0, 0.9)), and the cereal box brown or light-blue (RGB code (0.5, 0.7, 0.9)).

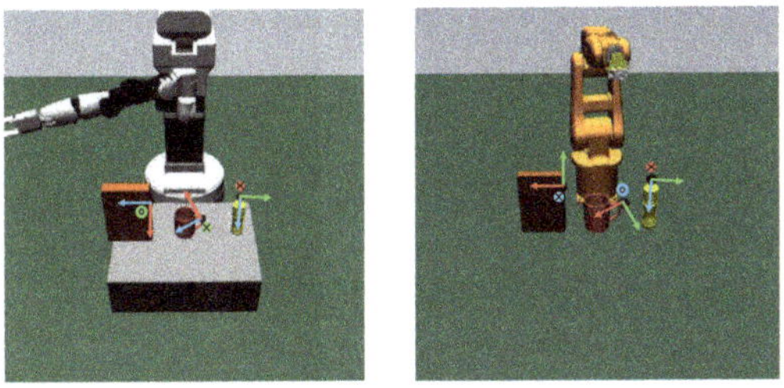

Figure 2. Environment observation with the three targets for the TIAGo and IRB120 scenarios. Note that we have added a black sphere on each grasping point for convenience. During training, the resolution is 64 × 64, and the sphere does not appear. The site axes correspond to the x coordinate in red, the y coordinate in green, and the z coordinate in blue.

3.3. MDP Definition

DRL problems can be defined as Markov Decision Processes (MDPs), which are characterized by a set of finite states, actions, a state transition probability matrix, a reward function, and the discount factor. Table 1 gathers the MDP definition of our problem. The state is a 64 × 64 × 3 RGB image captured at the beginning of each step. We use an episodic setup where each episode lasts 50 steps or less if the robot's gripper reaches the target within the rewarding distance and orientation. The reward given to the agent depends on the relative distance and orientation between the gripper and the grasping point. If the target is not reached, the agent receives a negative reward, whereas if it is, it receives a high positive value, as shown in Table 1, and the episode ends. The action set is formed by seven possible actions that scale the maximum joint movement range (Maximum Position Increment—MPI). At the beginning of the episode, the first and second joint orientations are set randomly within ±15% of their working range. Additionally, before starting the episode, one of the three targets is selected and placed randomly following a uniform distribution within the workspace area, which is a 20 × 25 cm rectangle for the TIAGo environment and a 60 × 20 cm rectangle for the IRB120 setup.

Table 1. MDP definition. $\mathrm{rel}_{\mathrm{dist}}$ and $\mathrm{rel}_{\mathrm{deg}}$ represent the relative distance and degrees between the gripper and the grasping point. MPI is the Maximum Position Increment. WRLL and WRUL are the Working Range Lower Limit and the Working Range Upper Limit of the robots. The target position is selected according to a uniform distribution.

MDP Definition	
Ep. Length	50 steps maximum
Reward configuration	$-2 * \mathrm{rel}_{\mathrm{dist}}^2 - \mathrm{rel}_{\mathrm{deg}}/70$ 100 if grasping point reached
Reward constraints	$\mathrm{rel}_{\mathrm{dist}} < 5\,\mathrm{cm}$ $\mathrm{rel}_{\mathrm{deg}} < 15^\circ$
Action set	0 ±MPI ±MPI/10 ±MPI/100
Initial robot config	$U(15\%\mathrm{WRLL}, -15\%\mathrm{WRUL})$ for joints 1 and 2
Initial target config	$\mathrm{x} \sim U(x_{min}, x_{max})$ $\mathrm{y} \sim U(y_{min}, y_{max})$

4. Methodology

4.1. Knowledge Graph and Embeddings

To provide external knowledge to the learner, we extracted contextual information of the scene from a knowledge graph $\mathcal{G}$. Then, we embedded the information in a latent space in the form of Knowledge Graph Embedding (KGE).

The objective of the KGE is to learn a continuous vector representation of a knowledge graph $\mathcal{G}$, which encodes vertices that represent entities $\mathcal{E}$ as a set of vectors $v_{\mathcal{E}} \in \mathbb{R}^{|\mathcal{E}| \times d_{\mathcal{E}}}$ (where $d_{\mathcal{E}}$ is the dimension of the vector of entities $\mathcal{E}$), and as a set of edges which represent relations $\mathcal{R}$ as mappings between vectors $W_{\mathcal{R}} \in \mathbb{R}^{|\mathcal{R}| \times d_{\mathcal{R}}}$, where $d_{\mathcal{R}}$ is the dimension of the vector of relations. The knowledge graph $\mathcal{G}$ is composed by triples (h, r, t), where $h, t \in \mathcal{E}$ are the head and tail of the relations, while $r \in \mathcal{R}$ is the relation itself. One example of such a triple is (*mug, hasColor, yellow*).

We initially used ANALOGY as a direct embedding model. Since we need to have a semantic representation of the scene of the agent, we defined a unified semantic embedding that is extracted by introducing an operator Γ that, given a graph $\mathcal{G}'$, selects the subgraph $\mathcal{S}$ from the larger knowledge graph $\mathcal{G}$ ($\mathcal{S} \subseteq \mathcal{G}$). This subgraph $\mathcal{S}$ contains all the nodes representing the objects that the agent is perceiving and the node at distance 1 from them. resulting in all relevant entities and their relationships pertinent to the specific scene the agent is interacting with. For instance, if the scene includes a mug, a bottle, and a cereal box, the subgraph $\mathcal{S}$ will contain nodes representing these objects and edges denoting their relationships with other nodes (e.g., (*mug, hasColor, red*), (*mug, isConnectedTo, handle*), (*bottle, hasShape, cylindrical*), (*cereal box, isMadeOf, cardboard*)). From the subgraph, a textual representation is obtained by concatenating the labels of the nodes and the edges in a single sentence. The sentence is then converted into a single embedding representing the scene. To effectively capture the semantic nuances of the scene, we replaced ANALOGY with a pre-trained embedding model, GloVe [35]. With GloVe, we generated a single embedding for the scene. The use of a pre-trained model on a large corpus ensures that the resulting vectors capture a wide range of semantic nuances. The dimension of these embeddings is set to 10×4 per word, allowing for a rich representation of the subgraph's content.

The resulting GloVe-based embeddings are then integrated into the learner's architecture. In our proposed setup, the embeddings are concatenated with the hidden activations from the visual observation layers before the policy approximation block. This strategic placement allows the DRL agent to utilize both the raw visual data and the rich contextual information provided by the embeddings, enhancing its ability to understand and interact with the environment.

4.2. Deep Reinforcement Learning Framework

4.2.1. Asynchronous Advantage Actor Critic (A3C)

For the sake of clarity, we first recap the concepts and terminology used to describe DRL. Model-free DRL includes value-based and policy-based approaches. The former are methods that learn either the estimation of the expected discounted cumulative reward, represented by the state-value function, $V^{\pi}(s)$, or the estimation of the expected discounted cumulative reward for taking action a in state s at step t, defined with the action-state value function, $Q^{\pi}(s, a)$ whose definitions are

$$V^{\pi}(s) = \mathbb{E}_{\pi}\left[\sum_{t=0}^{\infty} \gamma^{t} r_{t+1} \mid s_t = s\right] \tag{1}$$

$$Q^{\pi}(s, a) = \mathbb{E}_{\pi}\left[\sum_{t=0}^{\infty} \gamma^{t} r_{t+1} \mid s_t = s, a_t = a\right] \tag{2}$$

where $\mathbb{E}_\pi$ is the expectation following policy π, γ is the discount factor, r_{t+1} is the reward received at the time step $t+1$, and s_t is the current state.

The latter are algorithms seeking to maximize the discounted cumulative reward, but directly learning and optimizing the parametrized policy $\pi(s|a;\theta)$ [3].

A3C [36] is a policy-based method, with the particularity that the *actor* optimizes the policy, and, at the same time, the *critic* estimates the state-value function. A3C allows for the deployment of several agents that run in parallel, each with a unique environment instance, and asynchronously, they update a shared global network. This guarantees stable and efficient learning, reducing the correlation between the experience gathered by the instantiated agents. As shown in [37–39], A3C is a good alternative for discrete action spaces and high-dimensional state spaces.

Regarding the learning process, A3C uses the *Advantage Function*, $A(s,a) = Q(s,a) - V(s)$, to measure the goodness of an action. This function can be approximated by estimating the value of the current state, the reward, and the discounted value of the next state. This is the Temporal Difference error, or TD error at t:

$$\delta^{(t)} = r^{(t)} + \gamma V(s_{t+1};\theta) - V(s_t;\theta) \tag{3}$$

where $V(s_t;\theta)$ is the estimated value of s_t given the policy parameters θ. To smooth this estimation, we use the *General Advantage Estimator (GAE)*:

$$A_{GAE}^{(t)} = \gamma\lambda A_{GAE}^{(t+1)} + \delta^{(t)} \tag{4}$$

where $A_{GAE}^{(t)}$ is the GAE at time step t, and λ is the trace decay factor that controls the bias–variance trade-off. On the other hand, the loss is defined as the sum of the *Policy Loss* and the *Value Loss*, defined as

$$\mathrm{P_{loss}} = -\sum_i \Big(\log \pi(a_{ti}|s_{ti};\theta) \cdot A_{\mathrm{GAE}}^{(t)} - \beta H(\pi(s_t;\theta))\Big) \tag{5}$$

where i iterates over actions at t, $\pi(a_{ti}|s_{ti};\theta)$ is the probability of taking a_{ti}, β is the entropy regularization weight, and $H(\pi(s_t;\theta))$ is the policy entropy.

$$\mathrm{V_{loss}} = \frac{1}{2}\sum_t (R_t - V(s_t;\theta))^2 \tag{6}$$

where R_t is the n-step return at t.

4.2.2. Proposed Architecture

Figure 3 shows the proposed architecture. The state representation layers are defined by a convolutional layer CNN_1 with 3×3 kernel and stride 4, and another convolutional layer CNN_2 with 5×5 kernel and stride 2. Both layers have at the end a non-linear f implemented through the ReLU function, $f(x) = max(0,x)$. The policy approximation block is formed by a 1152×128 fully connected layer (FC), a Long Short-Term Memory (LSTM), whose capacity depends on the size of the integrated KGE. Since the agent commands the orientation of each robot joint, the A3C actor layers are n FC layers, where n is the number of robot joints, i.e., 7 for the TIAGo and 6 for the IRB120. The dimension of the n FC blocks is determined by the discrete set of actions, which has seven actions per robot joint (Table 1). We use the softmax function to convert the activations into probabilities. Therefore, the actor outputs are n 7-dimension arrays that contain the probability associated with each action. Finally, the A3C critic is a single FC layer that estimates the value of $V(s)$. We will refer to the baseline agent or baseline model (BM) as the agent trained under this architecture with just the RGB image. It will serve as a reference to quantify the improvement obtained by integrating the KGE.

The originality of our proposal lies in the integration of embeddings coming from a graph containing some environment semantic information. This new input, which does not change throughout the training, might be placed next to the policy approximation blocks, since it should complement the state representation from the "vision" layers. Therefore, the most appropriate place to integrate the KGE is before the LSTM layer, linking the embeddings to the activations coming from the FC. The FC layer receives the representation coming from the CNNs and provides a higher abstraction level. The size of the embeddings and, consequently, the size of the LSTM will depend on the case study. In general, the embedding concatenated is a 150-dimension array, except for the one used when DR is applied to the targets' color, which is a 300-dimension array. As a result, the LSTM size will be 128 when no KGE is integrated, 428 when the KGE includes all the information from the randomized targets, which is an assumable size for our architecture, and 278 otherwise.

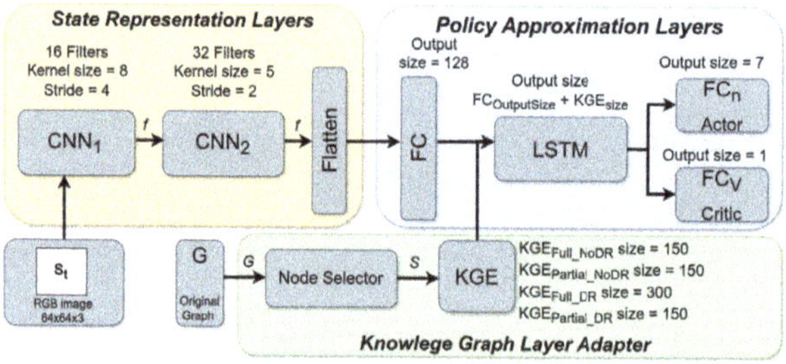

Figure 3. Proposed A3C architecture to integrate KGEs in the DRL agent's learning process. The KGE size depends on the number of features encoded. The output is divided into seven actors for the TIAGo and six for the IRB120, plus the critic value.

4.2.3. Training Procedure

We train the models for 70 million steps using the A3C algorithm. We use a random orthogonal initialization for all the layers. Based on the computational resources available, the number of asynchronous agents instantiated is seventeen plus one for the interim evaluation. This interim evaluation happens every 50,000 steps during training, and it assesses the global shared model for 40 episodes. These episodes keep the same 40 initial configurations, i.e., robot and target poses, across all the interim evaluations. At the beginning of each episode, a target is selected and placed randomly within the workspace area. The non-selected targets do not appear on the image. The rewarding distance is 5 cm, and the allowed orientation deviation is $\pm 15^{\circ}$.

The optimizer selected was the Root Mean Square Propagation (RMSprop) [40]. The values of the most relevant hyperparameters are learning rate $1e^{-4}$, discount factor (γ) 0.99, entropy weight (β) 0.01, trace decay (λ) 1, and RMSprop decay 0.99.

4.2.4. Post-Training Evaluation Procedure

After the training process, we select the best model according to the average return obtained in the interim evaluations. Then, we perform a post-training evaluation across 1000 episodes with a different seed and initial robot and target poses from the ones used during training. The rewarding distance and allowed orientation deviation in the post-training evaluation are 10 cm and $\pm 17^{\circ}$, respectively. We modify these training constraints because we believe that training in more restrictive circumstances yields better post-training results. In the assessment, we calculate the mean and standard deviation (Std) return, the mean and Std episode length, the mean, Std, and maximum failure distance, and the accuracy.

In our context, we understand accuracy as the percentage of successful episodes where the robot reached the target within the required distance. We formulate the accuracy as

$$\text{Accuracy (\%)} = \frac{\sum_{i=1}^{N} \mathbb{I}(\text{dist}_i \leq 10\,\text{cm})}{N} \times 100 \tag{7}$$

where N is the total number of evaluated episodes, $dist_i$ is the relative distance between the gripper and the target in the ith episode, and $\mathbb{I}$ is the indicator function.

4.2.5. Experiment Description

To analyze the influence of KGE in the learning process of a DRL agent, we present two sets of experiments. We first train three agents in an environment where each target has one possible color. The difference between these agents is the information we provide through the KGE. The BM agent has no KGE integrated. Then, there is an agent with only partial knowledge about the targets, just the object type (partial KGE). Finally, there is an agent with a complete description of the target type and color (full KGE).

In the second set of experiments, we decided to increase the complexity of the problem by introducing DR to the target color. Each object has two possible colors (Section 3) selected randomly at the episode beginning after the target choice and its placement. As in the first set, we train three different agents. A BM agent with no KGE, an agent with the partial KGE that contains only the object type, and an agent with the full KGE that has the object type and the possible colors per object.

We complement both groups of experiments with a quantitative and qualitative analysis of the joints' angles distribution to research how the exploration and exploitation capabilities of the agents might be affected by the KGEs.

5. Results

5.1. Experiments Without DR

Figures 4 and 5 show the training results in the environment with invariant features for the TIAGo and IRB120 robots, respectively. In the case of TIAGo, Figure 4 shows that the agent trained with the complete KGE learns faster and remains with a higher return than the agent trained with the partial KGE. In the first case, the best model is achieved at 48 M steps and has 72% accuracy, whereas in the second, it is at 36 M steps and has 70% accuracy. On the other hand, the differences with the BM in terms of learning speed and accuracy are higher. Although both agents reach the same steady regime, the best BM is at 60 M steps, and its accuracy is 60%. Hence, the improvement gain using KGE under a setup with invariant features is 12 percentual points.

Regarding the IRB120 robot, Figure 5 shows that, although there is a larger difference between the full KGE and the partial KGE trendlines, the accuracies of the best models are 92% and 91%, respectively. They are obtained in the 60 M step and the 59 M step. On the other hand, comparing the results with the BM, we notice that its learning is slower than the agent with the full KGE, which stands above the BM during the process. The performance of the BM best agent, which occurs in the 60 M step, reaches 80% accuracy. Therefore, the accuracy improvement in this setup without DR is also around 12% when using the KGE as an additional input to the model.

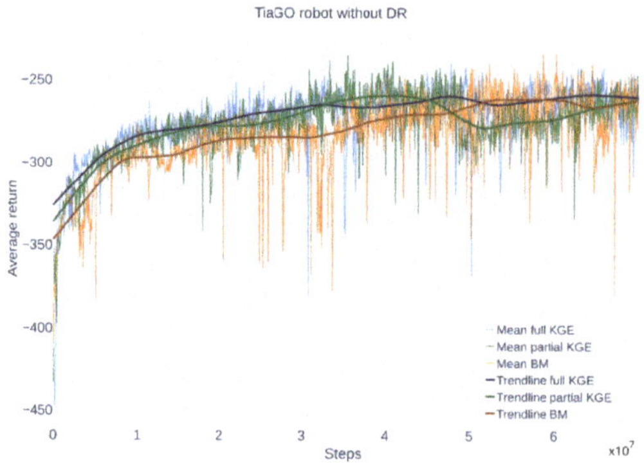

Figure 4. Training curves of the agents trained with the TIAGo and fixed targets' colors. The blue curve is the full KGE model, the green is the partial KGE, and the orange is the BM. The training process lasts for 70 M steps.

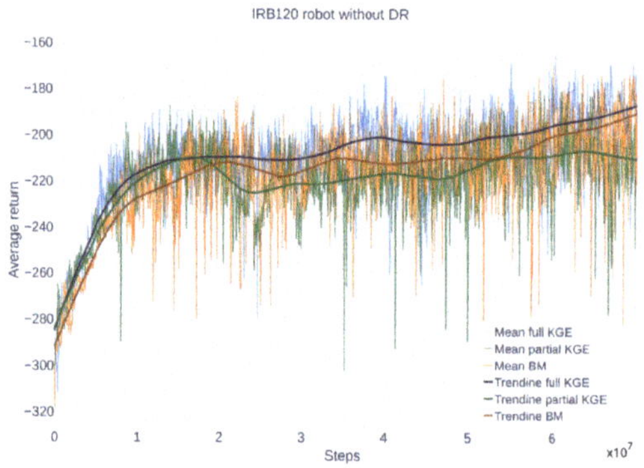

Figure 5. Training curves of the agents trained with the IRB120 and fixed targets' colors. The blue curve is the full KGE model, the green is the partial KGE, and the orange is the BM. The training process lasts for 70 M steps.

Lastly, the plots in the first column of Figure 6 show the joints' angles distribution for the TIAGo, while the Figure 6 second column graphs represent the IRB120 joints. We compare the distributions of the full KGE agents, represented in blue, with the BMs, shown in orange. The selected joints for the TIAGo are joints 3 and 4, and for the IRB120, they are joints 2 and 3. These joints were selected because they are the ones that present meaningful differences in the distribution of their values, which is interpreted as an improvement in the agent's ability to learn more straightforward the right policy. TIAGo's joint 3 experienced a slight reduction in the standard deviation of 0.01 points, but a significant mean variation. TIAGo's joint 4, however, has a large standard deviation decrease of 0.21 points and a mean shift. Moreover, the distribution changes to a bimodal shape, which might indicate some change in the agent's policy. In the case of IRB120, joint 2 has a standard deviation decrease of 0.02 points, as well as joint 3, which also presents a relevant mean difference in the distributions.

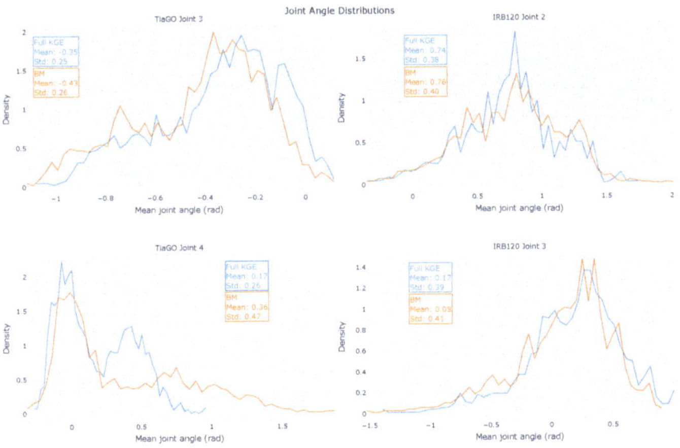

Figure 6. Joint angle distributions for the best full KGE agent (blue) and the best BM agent (orange), without DR. The first column corresponds to the TIAGo and the second to the IRB120. The selected joints for TIAGo are joints 3 and 4, and for the IRB120, they are joints 2 and 3.

5.2. Experiments with DR

Figures 7 and 8 present the training curves for the three models trained in the TIAGo and IRB120 setups when the targets' colors change along the training episodes. For the first scenario, we spot that the best model of the agent with the full KGE achieves 79% accuracy, higher than the 62% obtained by the agent with the partial KGE, and the 63% of the BM. Additionally, we can see that the learning is faster, and as a result, the best model is obtained at the 24 M step, earlier than in the partial KGE model, which happens at the 59 M step, and in the BM, at the 57 M step. Therefore, we see that the overall improvement of using a complete KGE with the target type and color in a setup where this last feature can change is around 16% with respect to the BM, and it is achieved in half of the training time.

In the scenario with the IRB120 robot, the agent with the complete KGE learns faster and yields higher than the BM. Whereas the first one reaches 86% accuracy in the 24 M step, the latter attains 76% in the 56 M step. If we compare the results with the model with partial KGE, it obtains 87% accuracy in the 43 M step, a comparable success rate. Table 2 collects the accuracy results of the trained agents.

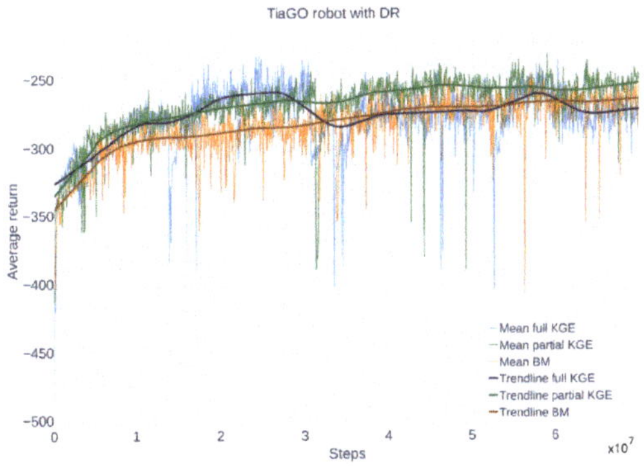

Figure 7. Training curves of the agents trained with the TIAGo and DR on the targets' colors. The blue curve is the full KGE model, the green is the partial KGE, and the orange is the BM. The training process lasts for 70 M steps.

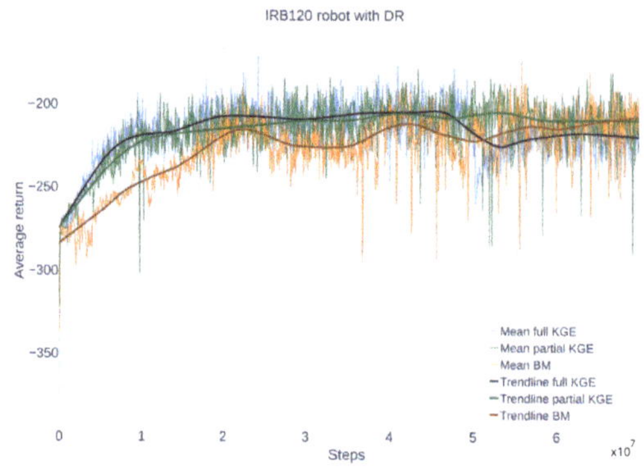

Figure 8. Training curves of the agents trained with the IRB120 and DR on the targets' colors. The blue curve is the full KGE model, the green is the partial KGE, and the orange is the BM. The training process lasts for 70 M steps.

Table 2. Experiments made to analyze the KGE influence in the DRL agent learning process, using TIAGo and IRB120 robotic arms. There are two sets: without DR and with DR. Each set includes three types of agents: BM, partial KGE, and full KGE.

Experiment Set	Robots	Agent Type	Accuracy %	Best Model Step
Without DR	TIAGo	BM	60%	60 M
		Partial KGE	70%	36 M
		Full KGE	72%	48 M
	IRB120	BM	80%	60 M
		Partial KGE	91%	59 M
		Full KGE	92%	60 M
With DR	TIAGo	BM	63%	57 M
		Partial KGE	62%	59 M
		Full KGE	79%	24 M
	IRB120	BM	76%	56 M
		Partial KGE	87%	43 M
		Full KGE	86%	24 M

Figure 9 exhibits the joints' angle distributions for the TIAGo and IRB120 models with DR. For the TIAGo, the most significant results appear on joints 2, 4, and 5. They are displayed in Figure 9's first column. In all of them, the full KGE agent, the blue curve, presents lower standard deviation values on the joints' distributions than the BM, the orange curve. In the case of joint 2, it decreases by 0.03 points, 0.04 points in joint 4, and 0.11 points in joint 5. For the IRB120, we analyze joints 1, 2, and 4, shown in Figure 9's second column. As in the TIAGo, the agent with the complete KGE, the blue plot, has less standard deviation than the BM, the orange curve. This reduction is of 0.05 points in joint 1, 0.07 points in joint 2, and 0.12 points in joint 4. We decided not to include the result of the other joints because it does not present a significant change.

Figures 10 and 11 illustrate the differences commented about the joint angle distribution depending on the presence or absence of semantic knowledge. As shown, the BM agent requires more intermediate movements from the beginning of the episode to reach the intermediate position where fine orientation and approach adjustments are made to grasp the target. In contrast, the agent equipped with the complete KGE typically reaches the grasping point more directly and with a more favorable pre-alignment, allowing for a smoother and more efficient approach. As Figure 11 shows, in steps 4 and 5 of the episode, the KGE agent has already reached the mug, and it is rotating the gripper to align it with the affordance point. Both episodes ended up successfully, but the BM agent needed 34 steps, while the full KGE agent needed only 20 steps.

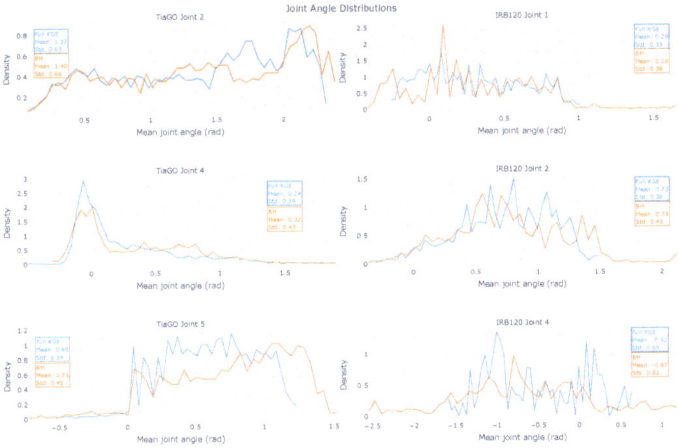

Figure 9. Joint angle distributions for the best full KGE agent (blue) and the best BM agent (orange), with DR. The first column corresponds to the TIAGo and the second to the IRB120. The selected joints for TIAGo are joints 2, 4, and 5, and for the IRB120, they are joints 1, 2, and 4.

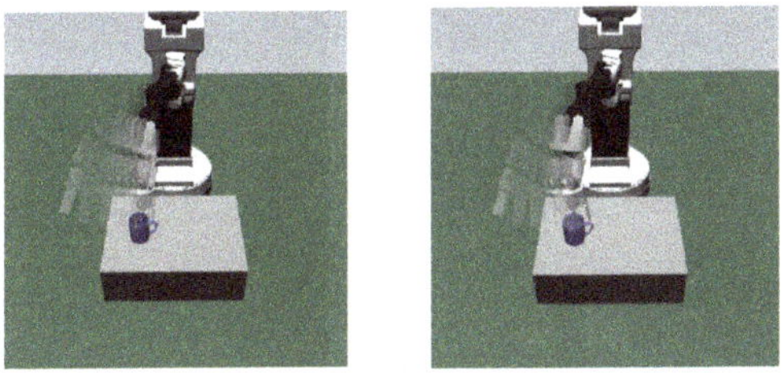

(**a**) (**b**)

Figure 10. TIAGo initial movements in the BM agent (**a**) and the full KGE agent (**b**) when DR is applied in the environment.

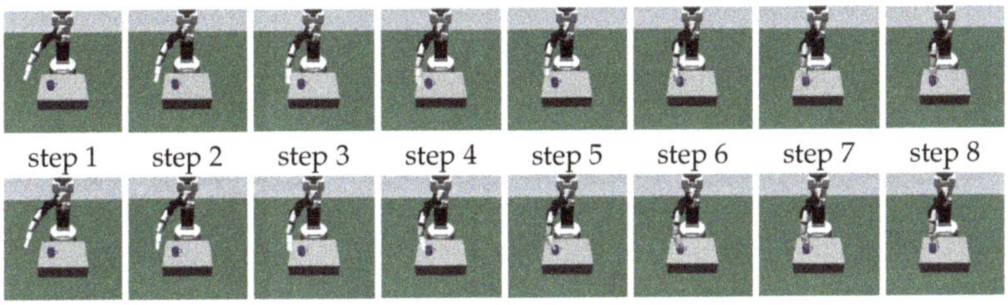

Figure 11. TIAGo initial movements in the BM agent (first row) and the full KGE agent (second row) when DR is applied in the environment. During the first steps of the episode, the KGE agent needs fewer steps to approach the object and reach a more suitable affordance orientation. Both episodes finished successfully, the BM agent in 34 steps and the full KGE agent in 20 steps.

5.3. Robustness of the Results

To statistically assess whether the observed differences in performance across models were significant, we conducted a one-way Analysis of Variance (ANOVA) on the success rates of each pair of models: TIAGo/IRB120 with DR or without DR for the full KGE agent and the BM agent. Each model was evaluated over 30 independent runs, each initialized with a different random seed, and each run consisted of 100 episodes. The null hypothesis (H_0) assumes that the mean success rates of the two models being compared are equal, i.e., any observed differences are due to random variation.

The results are gathered in Table 3. For the TIAGo robot, in the environment where DR is applied, the success rate of the full KGE agent was significantly higher than for the BM agent. Similarly, in the standard setting, the full KGE agent outperformed BM with statistical significance. For the IRB120 robot, the differences were also pronounced: under DR, the full KGE agent achieved significantly higher success rates than BM agent, and in the standard setting, the KGE agent again significantly outperformed the BM agent. These results confirm that the proposed KGE models consistently achieve superior performance compared to the BM across both robots and training conditions, with statistically significant margins. The high F-values and extremely low p-values indicate that the between-group variance greatly exceeds the within-group variance, supporting the robustness of the observed performance differences and justifying the rejection of the null hypothesis in all cases.

Table 3. One-way ANOVA results comparing KGE and BM agents for each robot and training setting (with and without domain randomization). Each result is based on 30 evaluations with different seeds, each consisting of 100 episodes.

Experiment Set	Robot	F-Value	p-Value
Without DR	TIAGo	16.47	<0.001
	IRB120	150.88	$<1 \times 10^{-17}$
With DR	TIAGo	247.14	$<1 \times 10^{-21}$
	IRB120	111.08	$<1 \times 10^{-14}$

6. Discussion

According to the results presented, we can say that, in general, the TIAGo's agents achieve less accuracy than IRB120's because their environment is more challenging. We argue that the morphology of the arm, the fact that it has one more DoF than the IRB120, its reach, and its color are factors that increase the complexity of the problem, although the task is the same.

In the experiments carried out with the TIAGo and with the IRB120 without DR, we consistently observed an improvement in the transitory regime and a decrease in the learning process time. For the TIAGo, there is an improvement of 25% in the learning time between the full KGE agent and the BM. In the IRB120, although the best models are achieved at the same time, the full KGE trendline is always above the BM, and its transitory is much faster. With respect to the success rates, in both setups, there is almost no difference between the full and the partial KGE. This might prove that, since the color is an invariant feature in these experiments, the presence of this specification on the embeddings does not change the agent performance significantly. However, we do obtain 12% accuracy improvement with respect to the BM when we evaluate the full KGE agent in the TIAGo and IRB120.

Regarding the analysis of the joints' angle distribution, we studied whether the use of KGEs in the learning process might involve an improvement in the exploration ability and, in consequence, better exploitation and more efficient learning. We see that in the TIAGo setup, there is a significant drop in the standard deviation value for joint 4 and a big change in the joint 3 mean. Conversely, in the IRB120 scenario, we did not obtain great standard deviation variations, but differences in the distribution means, as we see in joint 2 and, especially, in joint 3. This absence of meaningful changes might be in concordance with the fact that the BM is already close to mastering the task, so the differences in the exploration are not that relevant. These outcomes show that under the setup without DR, the exploration capability might present an enhancement, helping the KGE agents to achieve better policies sooner than the BM agents and, hence, learn faster.

When we apply DR to the targets' color, we see that for the TIAGo and the IRB120, the agents with the complete KGE are three times faster than the other agents, reducing the learning time needed by almost 60%. Moreover, their transitory regime is consistently better, as it happens in the experiments without DR. This might confirm that in a situation with an increase in the problem complexity resulting from varying features, adding contextual information helps in the learning process and the final performance. For the TIAGo, the full KGE agent performs 16% better than the partial KGE agent and the BM. The TIAGo's and IRB120's trendline analysis shows that the agent with the full KGE decreases its performance around the 30 M step and 50 M step, respectively, although later, they narrow the distance again with the other models. We argue that this might have happened due to a high learning rate and the fact that we do not use a learning rate decay mechanism.

For the IRB120, there is a 10% improvement between the full KGE agent and the BM. With this robot, it is noticeable that there is not a significant improvement between the full KGE and partial KGE success rate as it is in the TIAGo. We suggest that the color information does not involve an accuracy increase because of the scenario colors. Since the robot is orange, the RGB filters corresponding to the green and, especially, the blue channels do not have the same relevance as the red ones. Hence, this unbalance in the visual learned features might implicate an exclusion of part of the color information encoded in the KGE. This does not happen with the TIAGo because its arm is white, so it should equally use the filters for the three RGB channels.

Finally, in the outcomes obtained by analyzing the distribution of the joints' orientation, we noted a significant decrease in the standard deviation of the distribution for three of the TIAGo and IRB120 joints when KGEs are used. As in the previous set of experiments, this suggests that the agents can exploit the correct policies early thanks to the information encoded in the embeddings. This is consistent with the fact that these agents learn faster than the BM agents. This drop is larger and occurs in more joints than in the scenarios trained without DR, which might be because of the increase in the problem's complexity.

7. Limitations

While our experiments have shown improved learning efficiency in predefined environments, the reliance on predefined knowledge graphs may limit adaptability in dynamic and unstructured environments, where the robot must perform tasks that cannot benefit from semantic conditioning. The effectiveness of our approach depends on the completeness of the knowledge graph. If the graph lacks critical semantic relationships or contains incorrect relationships, the resulting embeddings may mislead the DRL agent, potentially degrading its performance. Our experiments were conducted in simulated environments to ensure controlled evaluations. However, since our observation is represented by an RGB image, transferring the learned policies to real-world robotic platforms would need an additional image-to-image translation method in order to efficiently bridge the simulation-to-reality gap.

8. Conclusions

In this paper, we demonstrate how adding contextual information about the environment, encoded in embeddings, helps DRL agents learn faster and more efficiently. These descriptions of the environment are obtained through the use of embeddings obtained by the selection of semantically relevant nodes in a knowledge graph. We have shown that adding these inputs to the DRL model does not significantly increase the computational time, and the impact on the model's architecture complexity is minimal. We propose an original architecture in which the embeddings are concatenated to the hidden activations of the FC layer that follows the representation learning blocks, thus complementing the visual information.

Overall, the results show an improvement in the agents' performance when KGEs are introduced. If the targets' colors are fixed, we noted a faster learning transitory regime, a decrease in the learning time, up to 25% in the TIAGo setup, and an increase in the agent's accuracy by 12% for both robots. If we apply DR to the targets' colors, the behavior of the transitory period remains better than the baseline, the decrease in the learning time rises up to 60% in both robots, and the agent's accuracy is increased up to 20% for TIAGo and 10% for IRB120. Additionally, we investigated the effect of KGEs on the joints' angle distribution to determine if there is an improvement in the agents' ability to exploit sooner the correct policy between the full KGE agent and the baseline.

For the experiments without DR, there are changes in the mean and standard deviation for some joints, but these are higher, and more joints are affected in the scenarios with DR. This suggests that, first, the use of KGEs enhances the agents' learning process and therefore, the learning time is reduced, the exploitation of the best policies happens earlier, and higher accuracies are achieved; and second, that providing the embedded information in a more complex problem leads to a more significant improvement than in simpler scenarios.

In conclusion, we have validated the proposed architecture that integrates contextual information about the environment, given as KGEs. We have proven that if DRL agents can leverage semantic knowledge, there is an improvement in their learning speed and success rate without significant increases in computational and architectural complexity.

Beyond the proposed application, the presented methodology shows promising capabilities in any scenario where DRL agents can leverage semantic knowledge sources. We leave for future work the analysis of how the agents' performance is affected depending on where the KGE input is placed; in our current approach, the placement was selected after the state representation layers, but different integration points within the network may influence how effectively the semantic information is combined with the learned visual features. Moreover, the next research steps also include the sim-to-real transfer for both setups, looking for a zero-shot or few-shot transfer where we can reuse, completely or partially, the knowledge learned in the virtual scenarios. To do so, we should explore domain adaptation techniques that minimize or potentially eliminate the need for real-world experience.

Author Contributions: Conceptualization, L.G.-L., V.S., J.B., Á.J.L.-L., and D.N.; methodology, L.G.-L., V.S., J.B., Á.J.L.-L., and D.N.; software, L.G.-L. and V.S.; validation, L.G.-L. and V.S.; formal analysis, L.G.-L. and V.S.; investigation, L.G.-L. and V.S.; resources, L.G.-L. and V.S.; data curation, L.G.-L. and V.S.; writing—original draft preparation, L.G.-L. and V.S.; writing—review and editing, L.G.-L., V.S., J.B., Á.J.L.-L., and D.N.; visualization, L.G.-L. and V.S.; supervision, J.B., Á.J.L.-L., and D.N. All authors have read and agreed to the published version of the manuscript.

Funding: This research received no external funding.

Data Availability Statement: Data will be available upon request.

Conflicts of Interest: The authors declare no conflicts of interest.

References

1. Lazaridis, A. Deep Reinforcement Learning: A State-of-the-Art Walkthrough. *J. Artif. Intell. Res.* **2020**, *69*, 1421–1471. [CrossRef]
2. François-Lavet, V.; Henderson, P.; Islam, R.; Bellemare, M.G.; Pineau, J. An introduction to deep reinforcement learning. *Found. Trends Mach. Learn.* **2018**, *11*, 219–354. [CrossRef]

3. Sutton, R.S.; Barto, A.G. *Reinforcement Learning: An Introduction*; MIT Press: Cambridge, MA, USA , 2018.
4. Goodfellow, I.; Bengio, Y.; Courville, A. *Deep Learning*; MIT Press: Cambridge, MA, USA, 2016.
5. Lecun, Y.; Bengio, Y.; Hinton, G. Deep learning. *Nature* **2015**, *521*, 436–444. [CrossRef] [PubMed]
6. Panzer, M.; Bender, B. Deep reinforcement learning in production systems: A systematic literature review. *Int. J. Prod. Res.* **2022**, *60*, 4316–4341. [CrossRef]
7. Rupprecht, T.; Wang, Y. A survey for deep reinforcement learning in markovian cyber–physical systems: Common problems and solutions. *Neural Netw.* **2022**, *153*, 13–36. [CrossRef]
8. Singh, B.; Kumar, R.; Singh, V.P. Reinforcement learning in robotic applications: A comprehensive survey. *Artif. Intell. Rev.* **2022**, *55*, 945–990. [CrossRef]
9. Qureshi, A.H.; Nakamura, Y.; Yoshikawa, Y.; Ishiguro, H. Intrinsically motivated reinforcement learning for human–robot interaction in the real-world. *Neural Netw.* **2018**, *107*, 23–33. [CrossRef]
10. Davidson, G.; Lake, B.M. Investigating Simple Object Representations in Model-Free Deep Reinforcement Learning. *Comput. Res. Repos.* **2020**. [CrossRef]
11. Kroemer, O.; Niekum, S.; Konidaris, G. A review of robot learning for manipulation: Challenges, representations, and algorithms. *J. Mach. Learn. Res.* **2021**, *22*, 1–82.
12. Cabi, S.; Colmenarejo, S.G.; Novikov, A.; Konyushkova, K.; Reed, S.E.; Jeong, R.; Zolna, K.; Aytar, Y.; Budden, D.; Vecerík, M.; et al. Scaling data-driven robotics with reward sketching and batch reinforcement learning. *Robot. Sci. Syst. XVI* **2019**. [CrossRef]
13. Wang, D.; Walters, R.; Zhu, X.; Platt, R. Equivariant *q* learning in spatial action spaces. In Proceedings of the 5th Conference on Robot Learning (CoRL). PMLR, Auckland, New Zealand, 14–18 December 2022; pp. 1713–1723.
14. Zeng, A.; Song, S.; Yu, K.T.; Donlon, E.; Hogan, F.R.; Bauza, M.; Ma, D.; Taylor, O.; Liu, M.; Romo, E.; et al. Robotic pick-and-place of novel objects in clutter with multi-affordance grasping and cross-domain image matching. *Int. J. Robot. Res.* **2022**, *41*, 690–705. [CrossRef]
15. Lee, Y.; Hu, E.S.; Lim, J.J. IKEA Furniture Assembly Environment for Long-Horizon Complex Manipulation Tasks. In Proceedings of the 2021 IEEE International Conference on Robotics and Automation (ICRA), Xi'an, China, 30 May–5 June 2021; pp. 6343–6349. [CrossRef]
16. Zhu, Y.; Wong, J.; Mandlekar, A.; Martín-Martín, R.; Joshi, A.; Nasiriany, S.; Zhu, Y. robosuite: A modular simulation framework and benchmark for robot learning. *arXiv* **2020**, arXiv:2009.12293. [CrossRef]
17. Delhaisse, B.; Rozo, L.; Caldwell, D.G. PyRoboLearn: A Python Framework for Robot Learning Practitioners. In Proceedings of the Conference on Robot Learning, PMLR, Virtual, 30 October–1 November 2020; Volume 100, pp. 1348–1358.
18. Ammanabrolu, P.; Hausknecht, M. Graph constrained reinforcement learning for natural language action spaces. *arXiv* **2020**, arXiv:2001.08837. [CrossRef]
19. Dambekodi, S.; Frazier, S.; Ammanabrolu, P.; Riedl, M.O. Playing text-based games with common sense. *arXiv* **2020**, arXiv:2012.02757. [CrossRef]
20. Pternea, M.; Singh, P.; Chakraborty, A.; Oruganti, Y.; Milletari, M.; Bapat, S.; Jiang, K. The RL/LLM Taxonomy Tree: Reviewing Synergies Between Reinforcement Learning and Large Language Models. *J. Artif. Int. Res.* **2024**, *80*. [CrossRef]
21. Wang, S.; Zhang, S.; Zhang, J.; Hu, R.; Li, X.; Zhang, T.; Li, J.; Wu, F.; Wang, G.; Hovy, E. Reinforcement Learning Enhanced LLMs: A Survey. *arXiv* **2025**, arXiv:2412.10400. [CrossRef]
22. Xie, T.; Zhao, S.; Wu, C.H.; Liu, Y.; Luo, Q.; Zhong, V.; Yang, Y.; Yu, T. Text2reward: Automated dense reward function generation for reinforcement learning. *arXiv* **2023**, arXiv:2309.11489. [CrossRef]
23. Carta, T.; Romac, C.; Wolf, T.; Lamprier, S.; Sigaud, O.; Oudeyer, P.Y. Grounding large language models in interactive environments with online reinforcement learning. In Proceedings of the 40th International Conference on Machine Learning. PMLR, Honolulu, HI, USA, 23–29 July 2023; pp. 3676–3713.
24. Ma, Y.J.; Liang, W.; Wang, G.; Huang, D.A.; Bastani, O.; Jayaraman, D.; Zhu, Y.; Fan, L.; Anandkumar, A. Eureka: Human-Level Reward Design via Coding Large Language Models. In Proceedings of the International Conference on Learning Representations (ICLR), Vienna, Austria, 7–11 May 2024. [CrossRef]
25. Quartey, B.; Shah, A.; Konidaris, G. Exploiting Contextual Structure to Generate Useful Auxiliary Tasks. *arXiv* **2023**, arXiv:2303.05038. [CrossRef]
26. Du, Y.; Watkins, O.; Wang, Z.; Colas, C.; Darrell, T.; Abbeel, P.; Gupta, A.; Andreas, J. Guiding Pretraining in Reinforcement Learning with Large Language Models. In Proceedings of the 40th International Conference on Machine Learning, PMLR, Honolulu, HI, USA, 23–29 July 2023; Volume 202, pp. 8657–8677.
27. Reid, M.; Yamada, Y.; Gu, S.S. Can wikipedia help offline reinforcement learning? *arXiv* **2022**, arXiv:2201.12122. [CrossRef]
28. Dasgupta, I.; Kaeser-Chen, C.; Marino, K.; Ahuja, A.; Babayan, S.; Hill, F.; Fergus, R. Collaborating with language models for embodied reasoning. *arXiv* **2023**, arXiv:2302.00763. [CrossRef]
29. Li, K.; Wang, J.; Yang, L.; Lu, C.; Dai, B. SemGrasp: Semantic grasp generation via language aligned discretization. In *Computer Vision—ECCV 2024*; Springer: Cham, Switzerland, 2025; pp. 109–127. [CrossRef]

30. Nie, M.; Chen, D.; Wang, D. Reinforcement learning on graphs: A survey. *IEEE Trans. Emerg. Top. Comput. Intell.* **2023**, *7*, 1065–1082. [CrossRef]
31. Miao, R.; Jia, Q.; Sun, F.; Chen, G.; Huang, H.; Miao, S. Semantic Representation of Robot Manipulation with Knowledge Graph. *Entropy* **2023**, *25*, 657. [CrossRef] [PubMed]
32. Pagès, J.; Marchionni, L.; Ferro, F. TIAGo: The Modular Robot That Adapts to Different Research Needs. Available online: https://www.clawar.org/wp-content/uploads/2016/10/P2.pdf (accessed on 22 June 2025).
33. ABB. *IRB 120: Product Specification*; ABB: Baden, Switzerland, 2023.
34. Todorov, E.; Erez, T.; Tassa, Y. MuJoCo: A physics engine for model-based control. In Proceedings of the IEEE International Conference on Intelligent Robots and Systems, Vilamoura-Algarve, Portugal, 7–12 October 2012; pp. 5026–5033. [CrossRef]
35. Pennington, J.; Socher, R.; Manning, C.D. Glove: Global vectors for word representation. In Proceedings of the 2014 Conference on Empirical Methods in Natural Language Processing (EMNLP), Doha, Qatar, 25–29 October 2014; pp. 1532–1543. [CrossRef]
36. Mnih, V.; Badia, A.P.; Mirza, M.; Graves, A.; Harley, T.; Lillicrap, T.P.; Silver, D.; Kavukcuoglu, K. Asynchronous methods for deep reinforcement learning. In Proceedings of the 33rd International Conference on International Conference on Machine Learning—Volume 48, New York, NY, USA, 20–22 June 2016; ICML'16, pp. 1928–1937. [CrossRef]
37. Gu, Z.; Jia, Z.; Choset, H. Adversary A3C for Robust Reinforcement Learning. In Proceedings of the International Conference on Learning Representations (ICLR), Vancouver, BC, Canada, 30 April 30–3 May 2018. [CrossRef]
38. Grondman, I.; Busoniu, L.; Lopes, G.A.; Babuška, R. A survey of actor-critic reinforcement learning: Standard and natural policy gradients. *IEEE Trans. Syst. Man Cybern. Part C Appl. Rev.* **2012**, *42*, 1291–1307. [CrossRef]
39. Babaeizadeh, M.; Frosio, I.; Tyree, S.; Clemons, J.; Kautz, J. Reinforcement Learning through Asynchronous Advantage Actor-Critic on a GPU. In Proceedings of the International Conference on Learning Representations (ICLR), Toulon, France, 24–26 April 2017. [CrossRef]
40. Tieleman, T.; Hinton, G. Lecture 6.5-rmsprop: Divide the gradient by a running average of its recent magnitude. *COURSERA Neural Netw. Mach. Learn.* **2012**, *4*, 26–31.

robotics

MDPI

Article

Safe Autonomous UAV Target-Tracking Under External Disturbance, Through Learned Control Barrier Functions

Promit Panja [1], Madan Mohan Rayguru [2,*] and Sabur Baidya [2,*]

[1] Department of Electrical and Computer Engineering, Virginia Tech, Blacksbrug, VA 24061, USA; ppanja@vt.edu
[2] Louisville Automation and Robotics Research Institute (LARRI), University of Louisville, Louisville, KY 40208, USA
* Correspondence: madanmohan.rayguru@louisville.edu (M.M.R.); sabur.baidya@louisville.edu (S.B.)

Abstract

Ensuring the safe operation of Unmanned Aerial Vehicles (UAVs) is crucial for both mission-critical and safety-critical tasks. In scenarios where UAVs must track airborne targets, they need to follow the target's path while maintaining a safe distance, even in the presence of unmodeled dynamics and environmental disturbances. This paper presents a novel collision avoidance strategy for dynamic quadrotor UAVs during target-tracking missions. We propose a safety controller that combines a learning-based Control Barrier Function (CBF) with standard sliding mode feedback. Our approach employs a neural network that learns the true CBF constraint, accounting for wind disturbances, while the sliding mode controller addresses unmodeled dynamics. This unified control law ensures safe leader-following behavior and precise trajectory tracking. By leveraging a learned CBF, the controller offers improved adaptability to complex and unpredictable environments, enhancing both the safety and robustness of the system. The effectiveness of our proposed method is demonstrated through the AirSim platform using the PX4 flight controller.

Keywords: UAV traget tracking; Control Barrier Function (CBF); safe control

1. Introduction and Related Works

The majority of the systems that drive our daily lives whether it be robotic systems, transportation systems, or manufacturing systems are of the Cyber-Physical nature. These applications couple control, computation, and physical dynamics into one integrated system, and most of the time these systems need to be designed with safety requirements in mind, hence becoming safety-critical [1–3]. Unmanned Aerial Vehicles (UAVs) have emerged as versatile aerial robotic platforms in the realm of cyber-physical systems [4] with applications spanning various domains, including surveillance, warfare, search and exploration in cluttered environments [5], agriculture, environmental monitoring, and even package delivery [4]. One of the key challenges in deploying UAVs within these contexts is ensuring safe operation, particularly in scenarios involving tracking and following dynamic targets like other UAVs. The safe operation of UAVs is crucial to prevent collisions that can cause damage to the vehicles, property, and people, especially in complex or congested environments. Collision avoidance systems enable UAVs to detect obstacles and autonomously adjust their flight paths, ensuring mission success and protecting both assets and humans. Collision avoidance is a paramount concern, as it directly influences the safety and effectiveness of the UAVs deployed in such an environment. In cooperative and coordinated autonomous operations with multiple UAVs, collisions can be controlled and avoided with more certainty, as we have full access to the control parameters of all the UAVs and can perform deterministic operations, e.g., coordinated UAV swarm, platooning, etc. However, when uncertainty is involved in some UAVs, and the platooning or swarm operation is not cooperative, ensuring collision avoidance is non-trivial.

Modern quadrotor unmanned aerial vehicles (UAVs) are becoming more and more popular, which increases the possibility that they will be used for malevolent purposes. One such example is an unregistered UAV flying in restricted airspace, occasionally moving erratically. Dynamic UAVs can be used to monitor and follow harmful targets under tight safety limitations in order to avoid these situations. However because the target's movement can be unpredictable, erratic, and random, the follower UAV must decide quickly and in real-time while still making sure it can follow the target safely [6].

Different collision avoidance techniques have previously been researched in the field of robotics. A classic technique is the dynamic window-based approach for obstacle avoidance [7], or the mixed-integer Linear Program-based controllers for safe online trajectory planning of unmanned vehicles [8]. A modern robotics approach to tracking and collision avoidance is by using state-of-the-art artificial intelligence techniques like Deep Neural Networks (DNN) and Reinforcement Learning (RL) [9–11]. A similar approach is also taken in [12] for implementing platooning control using the YOLO detector [13].

Classic collision avoidance methods like dynamic window approaches and mixed-integer linear programming can be computationally intensive, less scalable, and sensitive to model accuracy and sensor noise. AI-based techniques such as Deep Neural Networks and Reinforcement Learning offer adaptability but require large training data, high computation, and often lack formal safety guarantees. Geometric and potential field methods may suffer from local minima or overly conservative maneuvers, limiting efficiency in complex environments. A state-of-the-art nonlinear safety-critical control technique is *Control Barrier Function (CBF)* [1]. There has been numerous research conducted on collision avoidance in drones using CBF [4,6,14–18]. Drone platooning and swarm formation can also be achieved using CBFs as shown in [19–22]. CBFs are also useful in exploration in cluttered environments, as shown in [5,23]. Among other applications, CBFs have also been utilized for surveillance control for drone networks as shown in [24].

Various versions of CBFs like high order and adaptive CBFs have been shown to perform well in systems with high relative degree and uncertainty [17,25–30]. More recently modern data-driven and learning-based methods have been developed for finding CBFs with uncertainties [31–33], this allows the utilization of experimental data to find a CBF rather than manually formulating one which can be extremely difficult in practice. Estimators like neural networks and reinforcement learning have shown great success in estimating CBFs with uncertainties and input constraints [18,34–36].

Contribution. The main idea of our work is to design a robust safety controller for a quadrotor UAV that has to track and safely follow an unknown target UAV under various uncertainties in the form of unmodeled dynamics and external disturbances like wind. We summarize our contributions as follows:

- We introduce an offline, neural network-powered approach tailored for UAVs, to learn Control Barrier Function (CBF) constraints. This empowers the system to adeptly manage unmodeled dynamics and external disturbances prevalent in UAV flight.
- The proposed CBF is combined with a Sliding Mode Controller (SMC) to achieve robustness against model uncertainties. This ensures safe and reliable tracking even when the UAV's behavior deviates from its nominal model.
- We thoroughly evaluate the effectiveness of our proposed technique through an AirSim platform enabled with the PX4 controller. The results demonstrate its ability

to maintain safe target tracking despite external disturbances and modeling errors, which is suitable in real-time applications.

The paper is organized as follows. We introduce the mathematical model and formulate the problem in Section 2. Section 3 discusses the sliding mode control law and the proposed CBF method is presented in Section 4. The experimental results are presented in Section 5 followed by the conclusion.

2. System Dynamics and Problem Formulation

We consider the translational dynamics of a standard UAV for our work [37,38]. The position of the UAV evolves as:

$$\dot{p} = v$$

where $p = [x, y, z]^T$ is the position vector in 3 dimensions and v is the corresponding velocity vector. As we are focusing on position tracking of the target UAV, we can neglect the tilting-like effects on the body of the UAV. Ignoring the effect of drag and other rotational forces, the velocity of the UAV is influenced by the following factors.

- Gravitational effect along the negative z direction: $-g\hat{k}$, where g is gravitational constant.
- Thrust vector in inertial frame: $\frac{1}{m}\mathbf{R}T_b$, where m is the mass of UAV and $\mathbf{R}$ is the rotation matrix
- Lumped term for unmodelled dynamical effects (drag forces) and external disturbances in three dimensions: η

Considering these factors, we can express the velocity dynamics as:

$$\dot{v} = -g\hat{z} + \frac{1}{m}\mathbf{R}T_b + \frac{1}{m}\eta \tag{1}$$

where:

- $\hat{z} = \begin{bmatrix} 0 & 0 & 1 \end{bmatrix}^\top$ is the unit vector in the z-direction.
- $T_b = \begin{bmatrix} T_x & T_y & T_z \end{bmatrix}^\top$ represent thrust vector in body frame.
- $\eta = \begin{bmatrix} \eta_x & \eta_y & \eta_z \end{bmatrix}^\top$ is the lumped uncertainty.

The complete mathematical motion model for UAV considered in this paper can be expressed as:

$$\begin{aligned} \dot{p} &= vs. \\ \dot{v} &= \begin{bmatrix} 0 \\ 0 \\ -g \end{bmatrix} + \frac{1}{m}\mathbf{R}\begin{bmatrix} T_x \\ T_y \\ T_z \end{bmatrix} + \frac{1}{m}\eta \end{aligned} \tag{2}$$

Problem Definition

We consider a system of two UAVs, one is the target/leader UAV which flies in an unknown trajectory, and the other UAV called the follower has to track and follow the target while maintaining a safe distance from it at all times. Formally we can define this problem through the following safety condition.

$$||\mathbf{p}_l - \mathbf{p}_f|| - d_{safe} \geq 0 \tag{3}$$

Here, $\mathbf{p}_l \in \mathbb{R}^3$ and $\mathbf{p}_f \in \mathbb{R}^3$ are the positions of the leader and follower, respectively, and d_{safe} is the minimum distance that needs to be maintained to be considered safe. The ability to maintain this safety constraint at all times under unknown disturbances motivates us to build a safety controller for the follower UAV.

3. Sliding Mode Control Law

In this section, we derive a sliding mode control law, which does not have a barrier certificate. We will present the modifications for a learned CBF in the next section. Given a desired position p_d and desired velocity v_d, we consider a PID sliding surface s as:

$$s(t) = e_v(t) + \lambda_1 e_p(t) + \lambda_2 \int_0^t e_p(\tau)d\tau$$

where: $e_p = p - p_d$ is the position tracking error, $e_v = vs. - v_d$ is the velocity tracking error, λ_1, λ_2 are positive tuning gains, $\int_0^t e_p(\tau)d\tau$ is the integral of the position error.

The time derivative of the sliding surface can be derived as:

$$\dot{s}(t) = \dot{e}_v(t) + \lambda_1 \dot{e}_p(t) + \lambda_2 e_p(t)$$

Substitute $\dot{e}_v = \dot{v} - \dot{v}_d$ and $\dot{e}_p = vs. - v_d = e_v$:

$$\dot{s}(t) = (\dot{v} - \dot{v}_d) + \lambda_1 e_v + \lambda_2 e_p$$

Using the UAV dynamics, we can obtain (substitute the dynamics of $\dot{v}$ into $\dot{s}$):

$$\dot{s}(t) = \left(\begin{bmatrix} 0 \\ 0 \\ -g \end{bmatrix} + \frac{1}{m}\mathbf{R} \begin{bmatrix} T_x \\ T_y \\ T_z \end{bmatrix} + \frac{1}{m} d_{ext} \right) - \dot{v}_d + \lambda_1 e_v + \lambda_2 e_p$$

Using a standard Lyapunov function ($\frac{1}{2}s^2$), the control input $\begin{bmatrix} T_x & T_y & T_z \end{bmatrix}^\top$ must be designed to drive $\dot{s}(t) \leq 0$ for the state trajectories to be stable on the sliding surface. It's not impractical to assume that, the lumped uncertainty is upper bounded by:

$$\|\eta\| \leq \eta_{max}.$$

For a positive gain matrix K and a control input u_{smc}:

$$\begin{bmatrix} T_x \\ T_y \\ T_z \end{bmatrix} = m\mathbf{R}^{-1}(- \begin{bmatrix} 0 \\ 0 \\ -g \end{bmatrix} + \dot{v}_d - \lambda_1 e_v - \lambda_2 e_p - K\text{sgn}(s)), \tag{4}$$

one can prove that the time derivative of the Lyapunov function $s\dot{s}$ is negative definite, provided $K \geq \frac{1}{m}\eta_{max}$.

Note that the constants of the sliding surface λ_1, λ_2 should be positive scalars to make the surface $s = 0$ a suitable sliding surface (the state trajectories on the surface remain stable). If these constants are chosen to satisfy $\lambda_1^2 > 4\lambda_2$, then the errors e_p and e_v will not show oscillatory behavior on the surface and will converge to zero.

4. Learned CBF for External Wind Disturbances

We aim to design a safe controller for a UAV subjected to disturbances, ensuring the system remains within a predefined safe set.

4.1. Dynamical System Safety

Safety in dynamical systems refers to the system's ability to remain in a predefined safe set over time. Consider a system $\dot{x} = f(x)$ where $x \in \mathbb{R}^n$, and a safe set $\mathcal{S} = \{x \in \mathbb{R}^n : h(x) \geq 0\}$, defined by a smooth function $h : \mathbb{R}^n \to \mathbb{R}$. Nagumo's theorem states that the system remains safe if:

$$\dot{h}(x) \geq 0 \quad \forall x \in \partial\mathcal{S} \tag{5}$$

4.2. Control Barrier Function

For a nonlinear system:

$$\dot{x} = f(x) + g(x)u \tag{6}$$

a Control Barrier Function (CBF) $h(x)$ ensures that the system remains in the safe set $\mathcal{C}$ defined as,

$$\mathcal{C} = \{x \in \mathbb{R}^n : h(x) \geq 0\} \tag{7}$$

$$\partial\mathcal{C} = \{x \in \mathbb{R}^n : h(x) = 0\} \tag{8}$$

$$Int(\mathcal{C}) = \{x \in \mathbb{R}^n : h(x) > 0\} \tag{9}$$

by enforcing the following condition:

$$\frac{\partial h}{\partial x}(f(x) + g(x)u) \geq -\gamma(h(x)) \tag{10}$$

where γ is a class $\mathcal{K}$ function.

A Quadratic Program (QP) is used to enforce the CBF condition by filtering a potentially unsafe control input u_d from the feedback control law $k_d(x)$:

$$u^* = \underset{u}{\operatorname{argmin}} \|u - k_d(x)\|^2 \tag{11}$$

$$\text{s.t.} \quad \frac{\partial h}{\partial x}(f(x) + g(x)u) \geq -\gamma(h(x)) \tag{12}$$

This forms the basis of the CBF-QP controller, which will be compared with our proposed learned version in later sections.

4.3. Learned CBF

Drones operating under external disturbances such as wind or environmental factors often experience challenges in ensuring safety using traditional Control Barrier Functions (CBFs). To handle these uncertainties, we adapt the CBF framework to account for unknown disturbances using data-driven methods.

For the UAV dynamics, we modify the system to include an external disturbance $\mathbf{d}_{ext}(t)$:

$$\dot{\mathbf{v}} = \begin{bmatrix} 0 \\ 0 \\ -g \end{bmatrix} + \frac{1}{m}\mathbf{R}\begin{bmatrix} T_x \\ T_y \\ T_z \end{bmatrix} + \frac{1}{m}\mathbf{d}_{ext}(t) \tag{13}$$

where $\mathbf{d}_{ext}(t)$ represents external forces like wind disturbances. For our system defined with the safety condition (3) we choose an appropriate CBF candidate as,

$$h(\mathbf{p}, \mathbf{v}) := \|\mathbf{p}_l - \mathbf{p}_f\| - d_{safe} \tag{14}$$

Given the CBF $h(\mathbf{p}, \mathbf{v})$ defined on the safe set $\mathcal{C}$ (9), the CBF derivative can be written as:

$$\dot{h}(\mathbf{p}, \mathbf{v}, \mathbf{u}) = \dot{h}_b(\mathbf{p}, \mathbf{v}, \mathbf{u}) + \frac{\partial h}{\partial \mathbf{v}} \cdot \frac{1}{m}\mathbf{d}_{ext}(t) \tag{15}$$

Here, $\dot{h}_b(\mathbf{p}, \mathbf{v}, \mathbf{u})$ represents the baseline CBF derivative without disturbances:

$$\dot{h}_b(\mathbf{p}, \mathbf{v}, \mathbf{u}) = \frac{\partial h}{\partial \mathbf{p}} \cdot \mathbf{v} + \frac{\partial h}{\partial \mathbf{v}} \cdot \left(\begin{bmatrix} 0 \\ 0 \\ -g \end{bmatrix} + \frac{1}{m}\mathbf{R}\begin{bmatrix} T_x \\ T_y \\ T_z \end{bmatrix} \right) \tag{16}$$

To estimate the unknown disturbance effect, we use a neural network model $\mathcal{N}(\mathbf{p}, \mathbf{v}, \mathbf{u})$, which approximates:

$$\mathcal{N}(\mathbf{p}, \mathbf{v}, \mathbf{u}) \approx \frac{\partial h}{\partial \mathbf{v}} \cdot \frac{1}{m} \mathbf{d}_{ext}(t) \tag{17}$$

Thus, the complete estimated CBF derivative becomes:

$$\widehat{\dot{H}}(\mathbf{p}, \mathbf{v}, \mathbf{u}) = \dot{h}_b(\mathbf{p}, \mathbf{v}, \mathbf{u}) + \mathcal{N}(\mathbf{p}, \mathbf{v}, \mathbf{u}) \tag{18}$$

4.4. Training the Neural Network

To train the neural network, we collect data consisting of the state $\mathbf{p}_i$, $\mathbf{v}_i$, control input $\mathbf{u}_i$, and the true CBF derivative $\dot{h}_i$. The training target is computed as the difference between the true and baseline CBF derivatives:

$$y_i = \dot{h}_i - \dot{h}_b(\mathbf{p}_i, \mathbf{v}_i, \mathbf{u}_i) \tag{19}$$

The dataset for training the network becomes:

$$\mathcal{D}_{\text{train}} = \{(\mathbf{p}_i, \mathbf{v}_i, \mathbf{u}_i, y_i)\}_{i=1}^{N} \tag{20}$$

The neural network is trained to minimize the mean squared error (MSE) between the predicted disturbance effect and the true effect:

$$\min_{\mathcal{N}} \frac{1}{N} \sum_{i=1}^{N} (\mathcal{N}(\mathbf{p}_i, \mathbf{v}_i, \mathbf{u}_i) - y_i)^2 \tag{21}$$

4.5. Learned CBF-QP with Sliding Mode Control

Once trained, the neural network provides an estimate $\widehat{\dot{H}}(\mathbf{p}, \mathbf{v}, \mathbf{u})$, which incorporates external disturbances. This estimated CBF derivative can then be used in a CBF-QP combined with the sliding mode control law to ensure safety:

$$\mathbf{u}^* = \underset{\mathbf{u}}{\text{argmin}} \, \|\mathbf{u} - \mathbf{u}_{\text{smc}}\|^2 \tag{22}$$

$$\text{s.t.} \quad \widehat{\dot{H}}(\mathbf{p}, \mathbf{v}, \mathbf{u}) \geq -\gamma(h(\mathbf{p}, \mathbf{v})) \tag{23}$$

Here, $\mathbf{u}_{\text{smc}}$ is the sliding mode control input:

$$\mathbf{u}_{\text{smc}} = m\mathbf{R}^{-1} \left(\begin{bmatrix} 0 \\ 0 \\ -g \end{bmatrix} + \dot{\mathbf{v}}_d - \lambda_1 e_{\mathbf{v}} - \lambda_2 e_{\mathbf{p}} - K \, \text{sgn}(s) \right) \tag{24}$$

5. AirSim Experiments and Results

We opted for Microsoft AirSim [37,38] because AirSim provides an efficient simulation experience with its accurate physics engine and detailed sensor models, allowing for the replication of diverse sensor data under various environmental conditions. Furthermore, its compatibility with widely used robotics frameworks like ROS and PX4 (which are used in real-world UAVs) simplifies the transition from simulation to real-world deployment. AirSim is adaptable to real-world scenarios through its support for custom environments, sensors, and drone models, ensuring it meets the unique requirements of different projects. Moreover, AirSim has been extensively utilized for conducting high-fidelity simulation experiments by many researchers for various robotic platforms that were designed to deploy in the real world as demonstrated in these previous studies [39–43].

We conducted our experiments in the Microsoft AirSim simulator paired with PX4 [44,45] Autopilot in *Software In The Loop* mode as shown in Figure 1. The simulation setup was installed on a machine with AMD Ryzen 7 4700u APU, and running Windows 11. Microsoft AirSim v1.8.1 was installed with Unreal Engine v4.27 and paired with PX4 Autopilot v1.13.3.

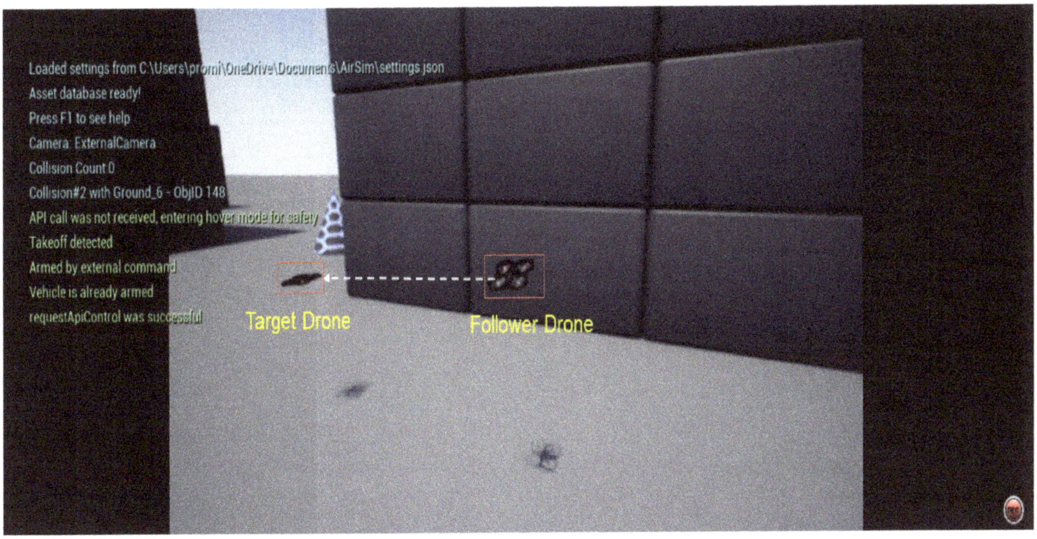

Figure 1. Simulation Setup in AirSim for two UAVs, one following the other.

The simulation environment is set with an unknown wind disturbance that is time-varying, which the follower drone does not have prior knowledge of. Another important thing that needs to be kept in mind is the follower drone also does not have prior knowledge of what trajectory the leader drone might take, the follower drone only has access to the translational states of the leader, i.e., the position and velocity which are updated in each control cycle which is roughly 10 Hz.

We constructed the neural network in Python 3.9 using the TensorFlow and Keras libraries, selecting an architecture with two fully connected hidden layers and ReLU activation functions for each hidden layer. ReLU was chosen due to its efficiency and effectiveness in capturing nonlinearity without suffering from vanishing gradients common to other activations such as sigmoid or tanh. The number of neurons per layer was determined empirically through preliminary experiments, balancing model capacity and the risk of overfitting, given the size of our dataset (20). The network was trained offline on an aggregated dataset collected from prior air sim experiments. Training utilized the Adam optimizer, a robust choice for deep networks due to its adaptive learning rates and efficient convergence properties. We applied mean squared error (MSE) as the loss function, reflecting the continuous nature of the CBF regression output. Model selection involved cross-validation on the training data, with a validation split set aside to monitor performance during training and tune hyperparameters. Performance assessment during training was based on validation loss. Further, after training, the model's generalization ability was evaluated on a separate held out test set.

AirSim has its own low-level PID controller for all the simulated vehicles it offers with the SimpleFlight class, but we decided to go with the open-source PX4 Autopilot flight stack as the low-level flight controller of choice, which takes the desired pitch ϕ_{des}, roll θ_{des}, and thrust T_{des} as the inputs from our proposed controller and sends the motor inputs to AirSim for our experiments. PX4 is a trusted open-source flight controller that is widely used for real-world UAVs and other aerial and ground vehicles. It also has great community support. PX4 can be used with a companion simulator either in *Software In The Loop* (SITL) or *Hardware In The Loop* (HITL) mode. For implementation, we need to calculate the desired roll (ϕ_{des}) and pitch (θ_{des}) using the filtered safe input (Equations (22)–(24)) $\mathbf{u}^* = \begin{bmatrix} T_x^{safe} & T_y^{safe} & T_z^{safe} \end{bmatrix}^\top$, which would be fed to the attitude controller along with the total thrust T^{safe} for generating motor inputs. We use the transformations shown in [46] to calculate the desired roll and pitch.

$$\phi_{\text{des}} = \frac{1}{mg}\left(T_x^{safe}\sin(\psi_{\text{des}}) - T_y^{safe}\cos(\psi_{\text{des}})\right)$$

$$\theta_{\text{des}} = \frac{1}{mg}\left(T_x^{safe}\cos(\psi_{\text{des}}) - T_y^{safe}\sin(\psi_{\text{des}})\right)$$

5.1. Results

To showcase and validate the performance of our proposed control strategy, we conducted several simulation experiments and divided them mainly based on the leader drone's trajectory— Straight Line and Circular Trajectory.

We conduct the experiments and compare the performance of four different types of controllers, which are, (a) Baseline-PDCBF which employs a PD control law for the linearized system along with the baseline CBF-QP (12). (b) Baseline-SMCBF which employs the sliding mode control law (24) along with the baseline CBF-QP. (c) Learned-PDCBF which employs the PD control along with learned CBF-QP (23), and similarly (d) Learned-SMCBF which employs the sliding mode control law along with learned CBF-QP. We chose $d_{safe} = 3\,\text{m}$.

5.2. Straight Line Trajectory

The plots in Figure 2a–d show the trajectory of the leader drone and the follower drone. In this scenario, the leader is flying in a straight line in the +ve x direction. We can see from the plots that all four controllers manage to track the leader drone but to understand if the controllers were able to maintain safety while tracking we need to analyze the CBF $h(\mathbf{x})$ vs. Time (t) plot.

In Figure 3, we can see that all four controllers perform pretty well up until time $t = 8$ s, after which the Baseline-PDCBF and Baseline-SMCBF violate the safety condition by going beyond the boundary of the safe set, that is the value of $h(\mathbf{x})$ drops below zero. We can also see that Baseline-PDCBF recovers after a while whereas Baseline-SMCBF continues to be in the unsafe region, the reason behind this is the unknown time-varying nature of the wind disturbance. Each simulation has a different wind condition which is completely random and is generated by the AirSim's wind model. In contrast, our proposed Learned-PDCBF and Learned-SMCBF controllers were able to maintain the safety condition throughout the experiment, with the latter performing slightly better with added robustness than the former.

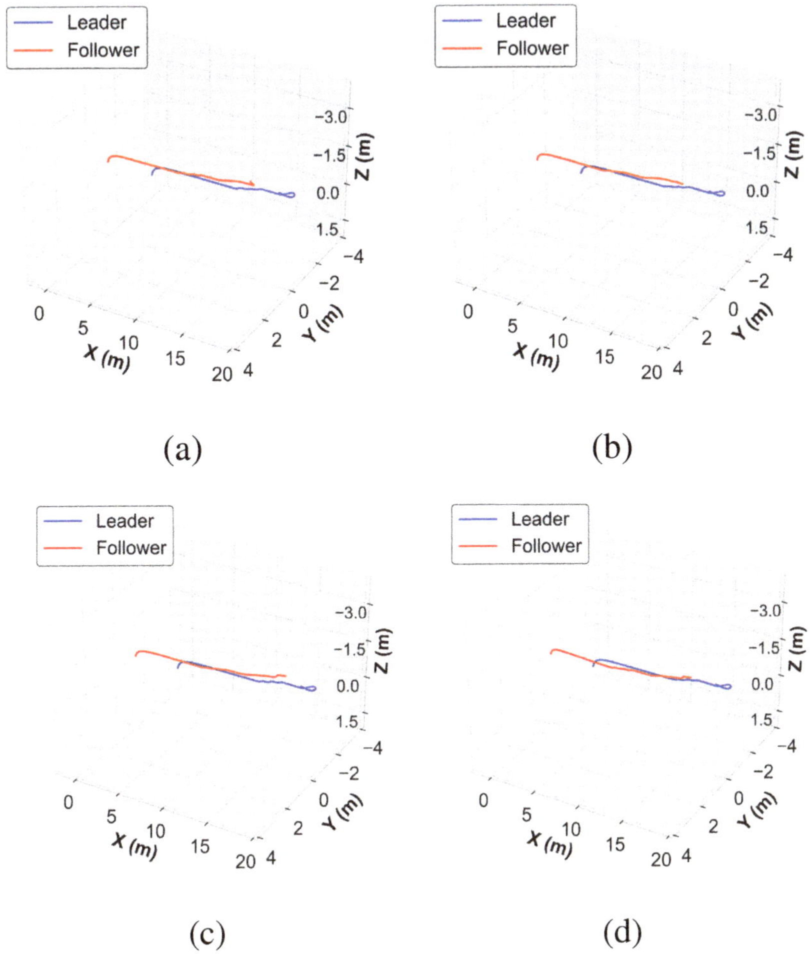

Figure 2. Straight-line trajectory tracking: (**a**) Baseline-PDCBF; (**b**) Learned-PDCBF; (**c**) Baseline-SMCBF; (**d**) Learned-SMCBF.

5.3. Circular Trajectory

Similarly, the plots in Figure 4a–d show the trajectories of the leader and the follower drone. Here the leader flies in a circular trajectory with a radius of 5.5 m in the clockwise direction starting from the center of the circle and stops after completing a full sweep. We can observe that all four controllers successfully track the leader drone, although from the plots it might look like the tracking performance is not good, we need to keep in mind that the follower drone is not just tracking the leader's position but also needs to maintain a safe distance while doing so. In this case to maintain the safety condition the controller forces the follower drone to drift inward maintain the safety condition.

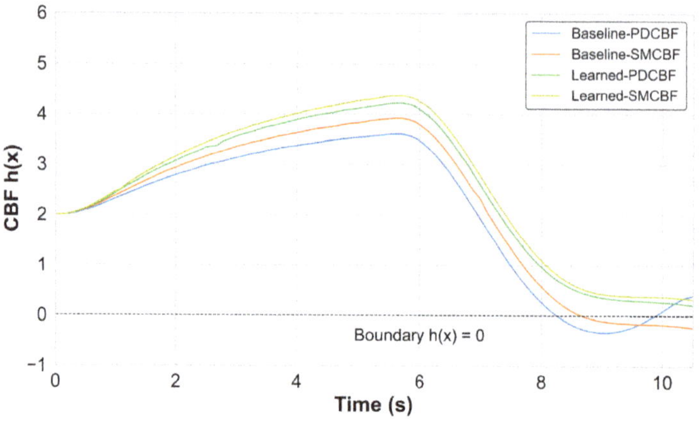

Figure 3. Value of $h(\mathbf{x})$ vs. Time for straight line trajectory.

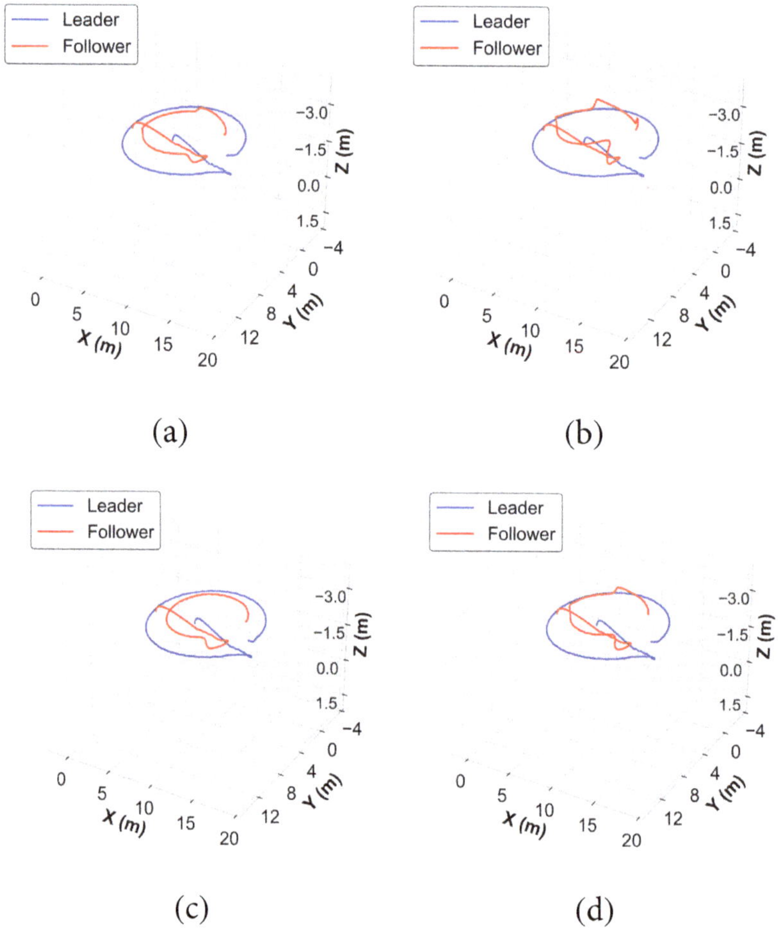

Figure 4. Circular trajectory tracking: (**a**) Baseline-PDCBF; (**b**) Learned-PDCBF; (**c**) Baseline-SMCBF; (**d**) Learned-SMCBF.

However similar to the previous scenario, not all the controllers were able to maintain safety at all times due to the presence of wind disturbance, namely, the Baseline-PDCBF and Baseline-SMCBF. By looking at the $h(\mathbf{x})$ vs. Time (t) plot in Figure 5 we can see that the Baseline-PDCBF and Baseline-SMCBF violate the safety condition approximately at $t = 7.5$ s, as the value of $h(\mathbf{x})$ goes below 0 for a brief period. Whereas our proposed controllers the Learned-PDCBF and Learned-SMCBF ensure safety all the time and successfully keep the system in the safe region.

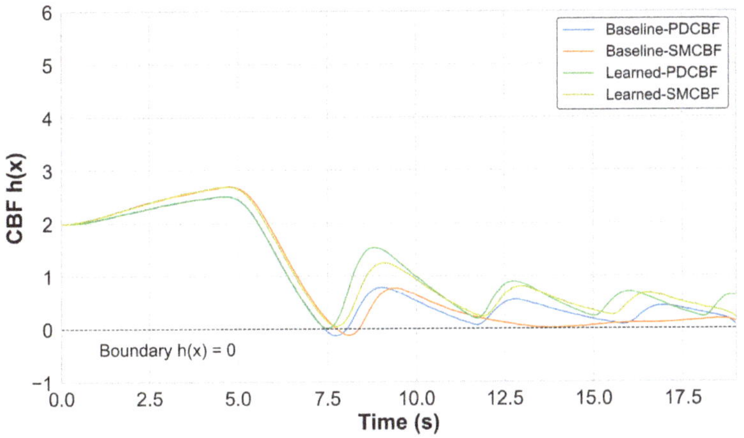

Figure 5. Value of $h(\mathbf{x})$ vs. Time for circular trajectory.

For the straight line scenario (Figure 2), the average RMSE between the leader and follower was 0.42 m for Baseline-PDCBF and 0.47 m for Baseline-SMCBF. Our proposed controllers—Learned-PDCBF and Learned-SMCBF—reduced this to 0.28 m and 0.30 m, respectively. Similarly, maximum tracking error decreased from around 1.0 m in the baseline cases to 0.6–0.65 m in the learned cases. For the circular trajectory (Figure 4), the RMSE increased slightly due to curvature, with baselines ranging from 0.55 to 0.60 m and learned controllers improving this to 0.37–0.39 m. Maximum tracking error also dropped from 1.2 m to 0.8 m with the learned approaches. The control framework remain unchanged as the number of UAVs increases. For multi-UAV or swarm scenarios, the method generalizes by applying the control barrier function (CBF) constraints between all relevant pairs or neighbors, supporting higher dimensional and collective behaviors.

6. Conclusions

The paper proposed a novel approach for ensuring safety in UAV pursuit scenarios, where the UAVs are operating under uncertainty like wind. We presented a learning-based Control Barrier Function (CBF) framework that effectively captures the impact of disturbances on the system, enabling accurate estimation of an accurate CBF constraint. Moreover, integrating the learning-based CBF with a sliding mode control law further enhances the system's robustness while upholding safety guarantees. The experimental results on AirSim demonstrate the effectiveness of our proposed method under varying wind disturbances. In contrast, the baseline controllers experienced safety violations in similar conditions. The combination of learning-based CBF and sliding mode control offers a robust solution for achieving both safety and reliability in UAV pursuit scenarios. Future research will focus on extending the framework to accommodate large-scale UAV swarms and dynamic environments with more complex disturbances, enhancing scalability and adaptability in real-world scenarios.

Author Contributions: Conceptualization, P.P., M.M.R. and S.B.; methodology, P.P.; software, P.P.; validation, M.M.R. and S.B.; formal analysis, M.M.R. and S.B.; investigation, M.M.R. and S.B.; resources, S.B.; data curation, P.P.; writing—original draft preparation, P.P.; writing—review and editing, M.M.R. and S.B.; visualization, P.P.; supervision, M.M.R. and S.B.; project administration, M.M.R. and S.B. All authors have read and agreed to the published version of the manuscript.

Funding: This research received no external funding.

Data Availability Statement: Data is contained within the article.

Conflicts of Interest: The authors declare no conflicts of interest.

References

1. Ames, A.D.; Xu, X.; Grizzle, J.W.; Tabuada, P. Control barrier function based quadratic programs for safety critical systems. *IEEE Trans. Autom. Control* **2016**, *62*, 3861–3876. [CrossRef]
2. Ames, A.D.; Coogan, S.; Egerstedt, M.; Notomista, G.; Sreenath, K.; Tabuada, P. Control barrier functions: Theory and applications. In Proceedings of the 2019 18th European Control Conference (ECC), Naples, Italy, 25–28 June 2019; IEEE: Piscataway, NJ, USA, 2019; pp. 3420–3431.
3. Panja, P. Survey Paper on Control Barrier Functions. *arXiv* **2024**, arXiv:2408.13271.
4. Xu, B.; Sreenath, K. Safe teleoperation of dynamic uavs through control barrier functions. In Proceedings of the 2018 IEEE International Conference on Robotics and Automation (ICRA), Brisbane, Australia, 21–25 May 2018; IEEE: Piscataway, NJ, USA, 2018; pp. 7848–7855.
5. Lerch, C.; Dong, D.; Abraham, I. Safety-critical ergodic exploration in cluttered environments via control barrier functions. In Proceedings of the 2023 IEEE International Conference on Robotics and Automation (ICRA), London, UK, 29 May–2 June 2023; IEEE: Piscataway, NJ, USA, 2023; pp. 10205–10211.
6. Panja, P.; Hoagg, J.B.; Baidya, S. Control Barrier Function Based UAV Safety Controller in Autonomous Airborne Tracking and Following Systems. *arXiv* **2023**, arXiv:2312.17215. [CrossRef]
7. Fox, D.; Burgard, W.; Thrun, S. The dynamic window approach to collision avoidance. *IEEE Robot. Autom. Mag.* **1997**, *4*, 23–33. [CrossRef]
8. Schouwenaars, T. Safe Trajectory Planning of Autonomous Vehicles. Ph.D. Thesis, Massachusetts Institute of Technology, Cambridge, MA, USA, 2006.
9. Zitar, R.A.; Mohsen, A.; Seghrouchni, A.E.; Barbaresco, F.; Al-Dmour, N.A. Intensive review of drones detection and tracking: Linear Kalman filter versus nonlinear regression, an analysis case. *Arch. Comput. Methods Eng.* **2023**, *30*, 2811–2830. [CrossRef]
10. Hsu, Y.H.; Gau, R.H. Reinforcement learning-based collision avoidance and optimal trajectory planning in UAV communication networks. *IEEE Trans. Mob. Comput.* **2020**, *21*, 306–320. [CrossRef]
11. Fraga-Lamas, P.; Ramos, L.; Mondéjar-Guerra, V.; Fernández-Caramés, T.M. A review on IoT deep learning UAV systems for autonomous obstacle detection and collision avoidance. *Remote Sens.* **2019**, *11*, 2144. [CrossRef]
12. Dai, X.; Nagahara, M. Platooning control of drones with real-time deep learning object detection. *Adv. Robot.* **2023**, *37*, 220–225. [CrossRef]
13. Redmon, J.; Divvala, S.; Girshick, R.; Farhadi, A. You only look once: Unified, real-time object detection. In Proceedings of the IEEE Conference on Computer Vision and Pattern Recognition, Las Vegas, NV, USA, 27–30 June 2016; pp. 779–788.
14. Tayal, M.; Kolathaya, S. Control Barrier Functions in Dynamic UAVs for Kinematic Obstacle Avoidance: A Collision Cone Approach. *arXiv* **2023**, arXiv:2303.15871.
15. Chen, Y.; Singletary, A.; Ames, A.D. Guaranteed obstacle avoidance for multi-robot operations with limited actuation: A control barrier function approach. *IEEE Control Syst. Lett.* **2020**, *5*, 127–132. [CrossRef]
16. Zhang, H.; Zhang, X.; Li, T.; Zhang, S.; Zhang, X. Barrier Function Enhanced Geometric Controller for Safe Control of a Quadrotor UAV. In Proceedings of the 2022 International Conference on Advanced Robotics and Mechatronics (ICARM), Guilin, China, 9–11 July 2022; IEEE: Piscataway, NJ, USA, 2022; pp. 187–192.
17. Wang, Z.; Hu, T.; Long, L. Multi-UAV safe collaborative transportation based on adaptive control barrier function. *IEEE Trans. Syst. Man Cybern. Syst.* **2023**, *53*, 6975–6983. [CrossRef]
18. Wang, L.; Theodorou, E.A.; Egerstedt, M. Safe learning of quadrotor dynamics using barrier certificates. In Proceedings of the 2018 IEEE International Conference on Robotics and Automation (ICRA), Brisbane, Australia, 21–25 May 2018; IEEE: Piscataway, NJ, USA, 2018; pp. 2460–2465.
19. Machida, M.; Ichien, M. Consensus-based control barrier function for swarm. In Proceedings of the 2021 IEEE International Conference on Robotics and Automation (ICRA), Xi'an, China, 30 May–5 June 2021; IEEE: Piscataway, NJ, USA, 2021; pp. 8623–8628.

20. Qing, W.; Chen, H.; Wang, X.; Yin, Y. Collision-free trajectory generation for UAV swarm formation rendezvous. In Proceedings of the 2021 IEEE International Conference on Robotics and Biomimetics (ROBIO), Sanya, China, 27–31 December 2021; IEEE: Piscataway, NJ, USA, 2021; pp. 1861–1867.
21. Gunnarsson, H.; Åsbrink, A. *Intelligent Drone Swarms: Motion Planning and Safe Collision Avoidance Control of Autonomous Drone Swarms*; Chalmers University of Technology: Gothenburg, Sweden, 2022.
22. Ghommam, J.; Luque-Vega, L.F.; Saad, M. Distance-based formation control for quadrotors with collision avoidance via Lyapunov barrier functions. *Int. J. Aerosp. Eng.* **2020**, *2020*, 2069631. [CrossRef]
23. Jang, I.; Kim, H.J. Safe Control for Navigation in Cluttered Space using Multiple Lyapunov-Based Control Barrier Functions. *IEEE Robot. Autom. Lett.* **2024**, *9*, 2056–2063. [CrossRef]
24. Dan, H.; Hatanaka, T.; Yamauchi, J.; Shimizu, T.; Fujita, M. Persistent object search and surveillance control with safety certificates for drone networks based on control barrier functions. *Front. Robot. AI* **2021**, *8*, 740460. [CrossRef]
25. Cohen, M.H.; Belta, C. High order robust adaptive control barrier functions and exponentially stabilizing adaptive control lyapunov functions. In Proceedings of the 2022 American Control Conference (ACC), Atlanta, GA, USA, 8–10 June 2022; IEEE: Piscataway, NJ, USA, 2022; pp. 2233–2238.
26. Lopez, B.T.; Slotine, J.J.E.; How, J.P. Robust adaptive control barrier functions: An adaptive and data-driven approach to safety. *IEEE Control Syst. Lett.* **2020**, *5*, 1031–1036. [CrossRef]
27. Taylor, A.J.; Ames, A.D. Adaptive safety with control barrier functions. In Proceedings of the 2020 American Control Conference (ACC), Denver, CO, USA, 1–3 July 2020; IEEE: Piscataway, NJ, USA, 2020; pp. 1399–1405.
28. Xiao, W.; Belta, C.; Cassandras, C.G. Adaptive control barrier functions. *IEEE Trans. Autom. Control* **2021**, *67*, 2267–2281. [CrossRef]
29. Liu, S.; Xiao, W.; Belta, C.A. Auxiliary-Variable Adaptive Control Barrier Functions for Safety Critical Systems. In Proceedings of the 2023 62th IEEE Conference on Decision and Control (CDC), Singapore, 13–15 December 2023.
30. Xiao, W.; Belta, C.; Cassandras, C. Adaptive Control Barrier Functions for Safety-Critical Systems. *arXiv* **2020**, arXiv:2002.04577. [CrossRef]
31. Tayal, M.; Zhang, H.; Jagtap, P.; Clark, A.; Kolathaya, S. Learning a Formally Verified Control Barrier Function in Stochastic Environment. *arXiv* **2024**, arXiv:2403.19332. [CrossRef]
32. Taylor, A.; Singletary, A.; Yue, Y.; Ames, A. Learning for safety-critical control with control barrier functions. In Proceedings of the Learning for Dynamics and Control. PMLR, Virtual, 10–11 June 2020; pp. 708–717.
33. Wu, Z.; Christofides, P.D. Control Lyapunov-barrier function-based predictive control of nonlinear processes using machine learning modeling. *Comput. Chem. Eng.* **2020**, *134*, 106706. [CrossRef]
34. Choi, J.; Castaneda, F.; Tomlin, C.J.; Sreenath, K. Reinforcement learning for safety-critical control under model uncertainty, using control lyapunov functions and control barrier functions. *arXiv* **2020**, arXiv:2004.07584. [CrossRef]
35. Cheng, R.; Khojasteh, M.J.; Ames, A.D.; Burdick, J.W. Safe multi-agent interaction through robust control barrier functions with learned uncertainties. In Proceedings of the 2020 59th IEEE Conference on Decision and Control (CDC), Jeju Island, Republic of Korea, 14–18 December 2020; IEEE: Piscataway, NJ, USA, 2020; pp. 777–783.
36. So, O.; Serlin, Z.; Mann, M.; Gonzales, J.; Rutledge, K.; Roy, N.; Fan, C. How to train your neural control barrier function: Learning safety filters for complex input-constrained systems. In Proceedings of the 2024 IEEE International Conference on Robotics and Automation (ICRA), Yokohama, Japan, 13–17 May 2024; IEEE: Piscataway, NJ, USA, 2024; pp. 11532–11539.
37. Microsoft AirSim. Available online: https://microsoft.github.io/AirSim/ (accessed on 14 September 2023).
38. Shah, S.; Dey, D.; Lovett, C.; Kapoor, A. AirSim: High-Fidelity Visual and Physical Simulation for Autonomous Vehicles. *arXiv* **2017**, arXiv:1705.05065.
39. Liu, B.; Wang, L.; Liu, M.; Xu, C.Z. Federated imitation learning: A novel framework for cloud robotic systems with heterogeneous sensor data. *IEEE Robot. Autom. Lett.* **2020**, *5*, 3509–3516. [CrossRef]
40. Liu, B.; Wang, L.; Chen, X.; Huang, L.; Han, D.; Xu, C.Z. Peer-assisted robotic learning: A data-driven collaborative learning approach for cloud robotic systems. In Proceedings of the 2021 IEEE international conference on robotics and automation (ICRA), Xi'an, China, 30 May–5 June 2021; IEEE: Piscataway, NJ, USA, 2021; pp. 4062–4070.
41. Ghods, R.; Durkin, W.J.; Schneider, J. Multi-agent active search using realistic depth-aware noise model. In Proceedings of the 2021 IEEE International Conference on Robotics and Automation (ICRA), Xi'an, China, 30 May–5 June 2021; IEEE: Piscataway, NJ, USA, 2021; pp. 9101–9108.
42. Baldini, F.; Anandkumar, A.; Murray, R.M. Learning pose estimation for UAV autonomous navigation and landing using visual-inertial sensor data. In Proceedings of the 2020 American Control Conference (ACC), Denver, CO, USA, 1–3 July 2020; IEEE: Piscataway, NJ, USA, 2020; pp. 2961–2966.
43. Saunders, J.; Saeedi, S.; Lil, W. Parallel reinforcement learning simulation for visual quadrotor navigation. In Proceedings of the 2023 IEEE International Conference on Robotics and Automation (ICRA), London, UK, 29 May–2 June 2023; IEEE: Piscataway, NJ, USA, 2023; pp. 1357–1363.

44. PX4 Autopilot. Available online: https://px4.io/software/software-overview/ (accessed on 14 September 2023).
45. Meier, L.; Honegger, D.; Pollefeys, M. PX4: A node-based multithreaded open source robotics framework for deeply embedded platforms. In Proceedings of the 2015 IEEE International Conference on Robotics and Automation (ICRA), Seattle, WA, USA, 26–30 May 2015; IEEE: Piscataway, NJ, USA, 2015; pp. 6235–6240. [CrossRef]
46. Michael, N.; Mellinger, D.; Lindsey, Q.; Kumar, V. The grasp multiple micro-uav testbed. *IEEE Robot. Autom. Mag.* **2010**, *17*, 56–65. [CrossRef]

robotics

MDPI

Article

Quadrupedal Locomotion with Passive Ventral Wheels: A Data-Driven Approach to Energy Efficiency Analysis

David Omar Al Tawil [1], Paolo Arena [1,*], Alessia Li Noce [1] and Luca Patanè [2]

[1] Department of Electrical, Electronic, and Computer Engineering (DIEEI), University of Catania, 95125 Catania, Italy; david.altawil@phd.unict.it (D.O.A.T.); alessia.linoce@unict.it (A.L.N.)
[2] Department of Engineering, University of Messina, 98122 Messina, Italy; lpatane@unime.it
* Correspondence: paolo.arena@unict.it

Abstract

In this paper, a hybrid locomotion approach is proposed and experimentally validated for a quadrupedal robot to enhance energy efficiency on mixed terrains. A mechanical solution was implemented by adding passive wheels on the robot's abdomen, to allow for gliding on flat portions of the faced terrains. This strategy aims to reduce the use of the legs, decreasing the overall energy consumption. To allow an efficient use of simulations, a data-driven approach was developed to estimate motor power consumption from joint dynamics on the real robot and subsequently applied within the simulation environment. The neural network achieved a coefficient of determination of $R^2 = 0.97$, ensuring accurate estimation of energy consumption under both simulated and real conditions. Experimental and simulated results show that the proposed sliding gait reduces the average Cost of Transport from approximately 4.5–6.0 during trotting to 0.8–1.1 during sliding, corresponding to a four–five-fold improvement in energy efficiency. Overall, the results demonstrate that a simple mechanical upgrade of the robot's body structure can significantly enhance locomotion efficiency and versatility on flat or slightly descending terrains.

Keywords: robotic locomotion; energy efficiency; cost of transport; quadrupedal robots; sliding gait

1. Introduction

Bio-inspired robotics offers a promising solution to the limitations of traditional robotic systems. By drawing inspiration from principles that nature has refined over millions of years of evolution [1], researchers have developed humanoid and quadrupedal robots that overcome many of the constraints of classical systems such as wheeled platforms. Terrestrial animals, for instance, have evolved diverse locomotion strategies that enable them to adapt to unpredictable terrain conditions. Legged locomotion allows for climbing steep, irregular surfaces and maintaining stability on soft or unsteady ground [2], explaining its evolutionary success. These natural solutions have, in turn, inspired roboticists to replicate similar levels of mobility and adaptability in artificial systems [3].

One of the main advantages of legged robots, especially quadrupeds, lies in their ability to traverse complex and uneven terrain with high adaptability and minimal environmental impact. Notable examples increasingly used in industrial, exploratory, and inspection contexts include MIT's Mini Cheetah, Boston Dynamics' Spot, and the Unitree family of robots [4–6]. Despite these advances, energy efficiency remains a major challenge, particularly for long-duration autonomous missions.

Previous studies have extensively examined the energetic performance of legged locomotion. Arena et al. [7] analysed bio-inspired control strategies on a Mini Cheetah-based platform, showing that adaptive gait modulation can reduce energy use on irregular terrain. Lei et al. [8] introduced specific resistance, also called Cost of Transport (COT), as a reference index for assessing quadrupedal energy efficiency, while Silva and Machado [9] demonstrated the influence of gait parameters and velocity COT values, typically ranging between 3 and 8. More recently, Zhu et al. [10] tackled the transferability gap between simulated and real-world energy performance through reinforcement learning. Collectively, these works highlight that legged robots remain considerably less efficient than wheeled systems (COT < 1), motivating the exploration of hybrid locomotion approaches combining walking and rolling.

Wheeled robots are characterised by their mechanical simplicity and high efficiency on flat, structured terrain. They are widely used in applications such as warehouse automation, delivery services, and urban inspection. Their main limitation, however, arises in unstructured or uneven environments, where their limited adaptability severely restricts mobility.

Some hybrid approaches, implemented, for instance, in the quadruped ANYmal, tried to enhance energy efficiency by integrating rolling mechanisms at the ends of the legs, thereby combining the advantages of walking and wheeled locomotion [11,12].

Similar strategies appear in other hybrid robots equipped with both legs and wheels, such as Centauro [13], Max [14], Pegasus [15], Cassino Hexapod [16], the hexapods from Brno University of Technology [17], and RoboSimian [18]. These platforms aim to improve mobility on flat terrain without compromising navigation capabilities on rough ground [19].

Such hybrid solutions enable greater efficiency and speed on structured surfaces while retaining adaptability on uneven terrain. However, they often suffer from limitations during transitions between locomotion modes and may compromise performance in traditional gaits by reducing grip during climbing. Moreover, versatile mobility introduces challenges such as complex control requirements, increased mechanical wear, and terrain-specific constraints. For instance, the transition from walking to skating requires sophisticated control to maintain stability and coordination [20].

For example, Chen et al. [21] proposed a quadruped with 4-DoF legs capable of roller-skating gaits such as swizzling, stroking, and trot-like motions. They reported a significant reduction in energy consumption compared with traditional trotting, based on mechanical energy computed from joint torque and velocity. However, their evaluation excluded electrical energy measurements or data-driven estimation, and the additional actuated joint increased mechanical complexity.

Our approach employs passive wheels mounted on the ventral side of the robot, enabling rolling locomotion when the terrain is sufficiently flat and thereby reducing actuator effort.

Accurately estimating the robot's total energy consumption is essential, since it depends not only on actuator torques but also on various onboard modules and sensors. Reliable data-driven estimation is, therefore, required to ensure consistency between simulation and real experiments.

To address this, an efficient estimator of total energy consumption based on real data is required to ensure the reliability of simulated evaluations. In this work, we propose a neural network-based reverse-engineering approach to energy estimation, designed to enhance the robustness of simulation campaigns.

In this paper, we present three main innovations that distinguish our approach from existing wheeled–legged robots:

- Passive wheels are mounted on the robot's belly rather than on the feet or tibia joints. This configuration preserves the robot's natural locomotion abilities (e.g., trotting and

crawling) while avoiding additional weight and inertial effects due to more complex limb mechanisms. During skating, redistributing part of the robot's weight onto the belly-mounted wheels reduces actuator load and smooths transitions between walking and gliding without compromising gait performance.
- A data-driven neural network model was implemented to estimate the robot's power consumption from joint torques and angular velocities. Trained on real-world data and integrated into the simulation environment, the model accurately predicts electrical current draw, thereby improving the realism of simulated dynamics. This enables the safe and reliable virtual testing of new gait patterns and control strategies, reducing experimental risk and accelerating development through more trustworthy simulations.
- A method is proposed to evaluate the energy efficiency of hybrid gaits in simulation using neural network-based power estimation, ensuring consistency with COT measurements obtained from real-world experiments.

2. Bio-Inspired Locomotion

Sliding-based locomotion is commonly observed in animals living in low-friction environments, where it provides an energy-efficient alternative to conventional walking.

One of the most notable examples is the emperor penguin, which engages in a behaviour known as tobogganing: gliding on its belly while propelling itself forward using its flippers and feet. This technique enables penguins to travel quickly and with reduced energy expenditure over long distances—often exceeding 100 km—between feeding grounds and breeding colonies. Their oil-coated feathers further reduce friction, improving overall gliding efficiency [22].

Otters have also been observed gliding on snow or muddy ground, possibly to conserve muscular energy and move more efficiently across slippery surfaces. Although sometimes interpreted as playful or social behaviour, this action also plays a role in locomotion efficiency under specific environmental conditions [23].

In broader biological studies, animals such as mudskippers and hatchling turtles demonstrate similar strategies that combine active propulsion with partial body–ground interaction. For example, hatchling turtles use their front flippers for thrust while keeping part of their body in contact with the substrate, thereby alternating between active movement and passive gliding [24].

Inspired by these natural strategies, this work aims to replicate comparable advantages in robotic locomotion. The focus lies not only on mimicking the motion patterns but also on adapting them to surfaces with low friction, such as snow or ice. To mimic these conditions, we equipped the robot with passive wheels mounted on its belly, as described in the following section. This biologically inspired approach represents the basis for the development of a gait that reduces actuator load and energy consumption by utilising passive gliding phases on suitable terrains. This natural pattern of alternating active and passive phases also inspired our hybrid gait, which alternates between muscular effort through leg activation and energy-saving sliding, similar to how some animals optimise their endurance during long journeys on certain terrains.

From a mechanical perspective, the design of the ventral wheel system followed an iterative process aimed at ensuring structural compatibility with the existing robot chassis. The wheels were positioned at the four corners of the main body, close to the natural contact line of the abdomen, to provide uniform support and minimise the elevation of the robot's centre of mass. The initial prototype featured three wheels (two at the rear and one at the front), but stability tests indicated the need for a fourth wheel to achieve balanced load distribution during motion. The vertical placement was adjusted to allow for free leg articulation while maintaining a posture similar to natural crouching, ensuring smooth transitions between walking and sliding gaits. The support structures were modelled on the torso geometry and 3D-printed to achieve a rigid, lightweight, and body-integrated configuration. Commercial castor wheels were selected to enable omnidirectional rolling and in-place turning, enhancing manoeuvrability in confined environments while keeping friction low. This design approach represents a balance between biological inspiration, efficient belly-supported sliding as observed in animals and the mechanical constraints typical of mobile robotic systems.

3. Control Framework

The proposed control architecture is composed of three interdependent modules, each responsible for a specific aspect of locomotion.

- Gait Generator: This module computes the foot contact configuration over time based on the desired linear and angular velocities along the vertical direction, ${v_z}^d$ and ${w_z}^d$, and the height of the centre of mass (COM) of the target body and outputs the desired foot position ${p_F}^d$ and velocity ${v_F}^d$ relative to the gait, as well as the phase delay between foot contacts. It also coordinates the sequence of leg movements by determining the start of the swing and stance phases for each foot and specifying their placement to achieve the commanded movement. In a sliding gait, the legs are synchronised in pairs in a caterpillar-like pattern (i.e. front left with front right and rear left with rear right) to ensure stability and avoid unwanted COM rotations that could be caused by other gaits that lack front-to-back synchrony. This coordination strategy reflects the biological inspiration underlying our hybrid gait: just as penguins and otters use synchronised sliding movements to maintain stability while reducing muscle effort on low-friction surfaces, the robot uses synchronisation of paired legs and passive abdominal support to achieve similar energetic benefits.
- Stance QP Controller: During the stance phase, the joint torques are calculated by solving a Quadratic Program (QP) that optimises the ground reaction forces under physical constraints such as friction cones. The dynamics of the four-legged robot are modelled using a floating-base rigid body abstraction, which takes into account both the effects of the active leg forces and the passive castor reactions. The equations of motion can be expressed as follows:

$$\mathbf{M}\,\mathbf{b} = \sum_{i=1}^{4} \mathbf{J}_i^{\top} \mathbf{f}_{\mathrm{leg},i} + \sum_{j=1}^{4} \mathbf{C}_j^{\top} \mathbf{f}_{\mathrm{castor},j} + \mathbf{f}_{ex} \tag{1}$$

where $\mathbf{M} \in \mathbb{R}^{6\times6}$ denotes the mass–inertia matrix of the floating base, $\mathbf{p}_{\mathrm{CoM}} \in \mathbb{R}^3$ is the position of the centre of mass in the body frame, and $\mathbf{b} = [\ddot{\mathbf{p}}_{\mathrm{CoM}}^{\top}; \dot{\omega}_b^{\top}]^{\top} \in \mathbb{R}^6$ is the generalised acceleration vector, comprising the linear acceleration of the centre of mass $\ddot{\mathbf{p}}_{\mathrm{CoM}} \in \mathbb{R}^3$ and the angular acceleration of the body $\dot{\omega}_b \in \mathbb{R}^3$. The matrices $\mathbf{J}_i \in \mathbb{R}^{6\times3}$ and $\mathbf{C}_j \in \mathbb{R}^{6\times3}$ represent the Jacobians mapping the forces applied at the i-th leg and j-th castor, respectively, to the wrench acting on the body, while $\mathbf{f}_{\mathrm{leg},i} \in \mathbb{R}^3$ and $\mathbf{f}_{\mathrm{castor},j} \in \mathbb{R}^3$ denote the contact forces at the legs and the passive castor reactions, and $\mathbf{f}_{ex} = [0\;0\;-mg\;0\;0\;0] \in \mathbb{R}^6$ accounts for the external translational forces and external torques acting on the floating base.
A QP problem is formulated to achieve an optimal distribution of contact forces that keeps the robot in equilibrium while respecting the contact and friction constraints. This QP explicitly considers the forces on the four legs of the robot, labelled $\mathbf{f}_{\mathrm{leg}}$, as well as on the passive ventral wheels, labelled $\mathbf{f}_{\mathrm{castor}}$. By transferring part of the robot's body weight to the wheels, the load on the legs is reduced during locomotion, while the robot can continue to generate the forces required for propulsion.

The mapping of contact forces to the base wrench is then defined as

$$\mathbf{A}_s \mathbf{f}_{\text{tot}} = \mathbf{b}_d, \tag{2}$$

where

$$\mathbf{A}_s = \begin{bmatrix} \mathbf{I}_3 & \cdots & \mathbf{I}_3 \\ \text{skew}(\mathbf{p}_1 - \mathbf{R} \cdot \mathbf{p}_{\text{CoM}}) & \cdots & \text{skew}(\mathbf{p}_8 - \mathbf{R} \cdot \mathbf{p}_{\text{CoM}}) \end{bmatrix}, \tag{3}$$

$$\mathbf{b}_d = \begin{bmatrix} m(\ddot{\mathbf{p}}^d_{\text{CoM}} - \mathbf{g}) \\ \mathbf{I}_b \dot{\omega}^{\text{d}} \end{bmatrix} \tag{4}$$

with $\mathbf{p}_i$ denoting the positions of legs ($i = 1 \dots 4$) and wheels ($i = 5 \dots 8$) in the world frame, $\mathbf{A_s} \in \mathbb{R}^{6x24}$, and the skew-symmetric operator skew$(\cdot)$ converting vectors into matrices to compute torque contributions. The vector $\mathbf{b}_d \in \mathbb{R}^6$ is the generalised vector of the desired acceleration. The matrix R is the rotation matrix that transforms vectors from the body frame to the global frame.
The vector of contact forces is structured as follows:

$$\mathbf{f}_{\text{tot}} = \begin{bmatrix} \mathbf{f}^T_{\text{leg},1} & \mathbf{f}^T_{\text{leg},2} & \mathbf{f}^T_{\text{leg},3} & \mathbf{f}^T_{\text{leg},4} & \mathbf{f}^T_{\text{castor},1} & \mathbf{f}^T_{\text{castor},2} & \mathbf{f}^T_{\text{castor},3} & \mathbf{f}^T_{\text{castor},4} \end{bmatrix}^T,$$

where $\mathbf{f}_{\text{leg},i} \in \mathbb{R}^3$, with $i = 1, \dots, 4$, represents the force at the i-th leg, while $\mathbf{f}_{\text{castor},j} \in \mathbb{R}^3$, with $j = 1, \dots, 4$ corresponds to the passive wheel forces. The passive wheels are included directly in $\mathbf{f}_{\text{tot}}$, with constraints enforcing zero horizontal force components and allowing for vertical reaction to support weight.
The sliding-force control problem is then posed as the following QP:

$$\begin{aligned} \min_{\mathbf{f}_{tot} \in \mathbb{R}^{24}} \quad & \frac{1}{2} \mathbf{f}_{tot}^\top \mathbf{G}_s \mathbf{f}_{tot} + \mathbf{g}_{0,s}^\top \mathbf{f}_{tot} \\ \text{s.t.} \quad & \mathbf{A}_s \mathbf{f} = \mathbf{b}_d, \\ & \mathbf{C}_E \mathbf{f}_{tot} = \mathbf{c}_E, \\ & \mathbf{C}_I \mathbf{f}_{tot} \leq \mathbf{c}_I. \end{aligned} \tag{5}$$

The cost matrix $\mathbf{G}_s \in \mathbb{R}^{24 \times 24}$ is composed of three terms:

$$\mathbf{G}_s = \mathbf{A}_s^\top \mathbf{S} \mathbf{A}_s + \alpha \mathbf{W} + \beta \mathbf{U}, \tag{6}$$

where $\mathbf{S} \in \mathbb{R}^{6 \times 6}$ penalises deviations from the desired wrench, $\mathbf{W} \in \mathbb{R}^{24 \times 24}$ penalises the magnitude of contact forces, and $\mathbf{U} \in \mathbb{R}^{24 \times 24}$ penalises temporal variations in these forces. The scalar parameters α and β act as regularisation coefficients, weighting the respective contributions of force magnitude and smoothness.
The linear term in the cost function is defined as

$$\mathbf{g}_0^\top = -\mathbf{b}_d^\top \mathbf{S} \mathbf{A_s} - \beta \mathbf{f}_{\text{prev}}^\top \mathbf{U} + \Delta \mathbf{f}_{\text{res}}^\top, \tag{7}$$

where $\mathbf{f}_{prev} \in \mathbb{R}^{24}$ contains the contact forces (legs and castor reactions) from the previous control step, and $\Delta \mathbf{f}_{res} \in \mathbb{R}^{24}$ represents fictitious resistive forces applied to the legs to compensate for undesired body motion:

$$\Delta \mathbf{f}_{\text{res}} = \begin{bmatrix} -\frac{1}{2}\mathbf{f}_{\text{res},1} \\ -\frac{1}{2}\mathbf{f}_{\text{res},2} \\ -\frac{1}{2}\mathbf{f}_{\text{res},3} \\ -\frac{1}{2}\mathbf{f}_{\text{res},4} \\ \mathbf{0}_{12} \end{bmatrix}, \quad \mathbf{f}_{\text{res},i} \in \mathbb{R}^3, \ i = 1,\ldots,4. \tag{8}$$

The resistive forces acting on the robot base are modelled as a combination of viscous and Coulomb friction. A pure Coulomb model leads to a discontinuity at zero velocity, which is undesirable for both numerical optimisation and simulation. To overcome this problem, a smoothed Coulomb formulation was used in [25], where the discontinuity of the sign function is regulated by the parameter ε:

$$\mathbf{f}_{\text{res}} = k_{\text{visc}}\,\mathbf{v}_{xy} + k_{\text{coul}}\frac{\mathbf{v}_{xy}}{\sqrt{\mathbf{v}_{xy}^\top \mathbf{v}_{xy} + \varepsilon^2}}, \tag{9}$$

where $\mathbf{v}_{xy} \in \mathbb{R}^2$ denotes the horizontal velocity of the robot base, and k_{visc}, k_{coul}, and ε are tunable parameters regulating viscous and smoothed Coulomb friction. The parameter values used in the optimisation problem are summarised in Table 1.

Table 1. Controller parameters for the sliding gait. The matrices **W** and **U** used in the optimisation problem are diagonal, with the vectors W_s and U_s being placed on their diagonals. The first 12 entries of W_s and U_s correspond to the leg forces, while the last 12 entries correspond to the castor forces. Formally, we have $\mathbf{W} = \text{diag}(W_s), \mathbf{U} = \text{diag}(U_s), W_s, U_s \in \mathbb{R}^{24}, W_s = [\,W_s^{\text{legs}}, W_s^{\text{castor}}\,], \quad U_s = [\,U_s^{\text{legs}}, U_s^{\text{castor}}\,]$.

Symbol	Control Parameter
$\mathbf{S}$	$\mathbf{S} = \text{diag}(20, 20, 50, 450, 450, 450)$
α	0.0001
β	0.2
k_{visc}	10.0
k_{coul}	500.0
ε	0.05
W_s	$\begin{bmatrix} W_s^{\text{legs}} & : & 10,10,4,\ 10,10,4,\ 10,10,4,\ 10,10,4 \\ W_s^{\text{castor}} & : & 0,0,8,\ 0,0,8,\ 0,0,8,\ 0,0,8 \end{bmatrix}$
U_s	$\begin{bmatrix} U_s^{\text{legs}} & : & 3,3,3,\ 3,3,3,\ 3,3,3,\ 3,3,3 \\ U_s^{\text{castor}} & : & 0,0,3,\ 0,0,3,\ 0,0,3,\ 0,0,3 \end{bmatrix}$

The standard quadratic programming formulation

$$\min_{\mathbf{x}} \frac{1}{2}\mathbf{x}^\top H\mathbf{x} + g^\top \mathbf{x} + c \tag{10}$$

includes the linear term $g^\top \mathbf{x}$, which captures any bias or affine contribution to the cost function, whereas the constant term c does not influence the minimisation process [26,27]. Therefore, any additive term that enters linearly with respect to the decision variables, such as a residual force bias $\Delta \mathbf{f}_{res}$, should be included in g. In our formulation, the setting

$$\mathbf{g} = -\mathbf{A}_s^\top \mathbf{S}\mathbf{b}_d - \beta\mathbf{U}\mathbf{f}_{prev} + \Delta\mathbf{f}_{res} \tag{11}$$

is thus formally correct and preserves the convexity of the QP.
The parameters reported in Table 1 were obtained through an iterative optimisation process aimed at minimising oscillations and accurately reproducing the real robot's dynamic response during sliding locomotion. Specifically, the control gains α and β were empirically adjusted to achieve stable locomotion in simulation. The diagonal entries of W_s and U_s are carefully tuned according to the role of each contact: active leg forces are weighted more uniformly to enforce consistent regulation across all legs, while passive castor wheel forces are assigned different weights to reflect their limited actuation, with zero penalties on unconstrained directions. This selection of values ensures that the optimisation produces contact forces that are both physically realistic and temporally smooth, providing stable and reliable performance during sliding locomotion. The viscous and Coulomb friction coefficients (k_{visc}, k_{coul}) were identified by matching the simulated and measured horizontal forces required to initiate motion, thereby reproducing the resistive behaviour of the real passive wheels. The smoothing term ε was set to a small value (0.05) to preserve numerical continuity around zero velocity in the smoothed Coulomb model.
The equality and inequality constraint matrices, $\mathbf{C}_E$ and $\mathbf{C}_I$, enforce the physical limits described in Table 2.

Table 2. Equality ($\mathbf{C}_E$) and inequality ($\mathbf{C}_I$) constraints for contact forces.

Symbol	Constraint	Formulation
$\mathbf{C}_E$	Non-contact legs	$\mathbf{f}_{leg,i} = 0$
$\mathbf{C}_E$	Castors (no horizontal force)	$f_x = 0, \;\; f_y = 0$
$\mathbf{C}_I$	Global vertical force limit	$\sum f_z \leq mg$
$\mathbf{C}_I$	Contact feet (friction cone)	$\mathbf{C}_{\text{fric}} \, \mathbf{R}_{\text{terrain}}^{\top} \mathbf{f}_{leg,i} \leq 0$

This formulation takes into account a slight castor slip during the sliding phases and ensures a robust and realistic interaction between the robot and the terrain. The resulting contact forces are then used to calculate the forward joint torques as follows:

$$\tau_{ff} = \mathbf{J}^{\top} \mathbf{f}_{leg} \tag{12}$$

- Swing Controller: The trajectories of the swing legs are generated using a cycloid profile that enables smooth acceleration and deceleration phases. The desired foot positions are tracked using inverse kinematics and regulated by a proportional–derivative (PD) controller in the joint space, which enables accurate and stable foot positioning, even during transitions between walking and sliding.

  Overall, the commanded torques to be sent to the robot are as follows:

$$\tau = \tau_{ff} + k_p \cdot (q^d - q) + k_d \cdot (dq^d - dq) \tag{13}$$

The overall control scheme and robot configuration are shown in Figures 1 and 2. It operates at the constant frequency of 500 Hz, locked by a specifically designed timing logic, which guarantees the performance of the global control loop, similarly to that adopted in [28].

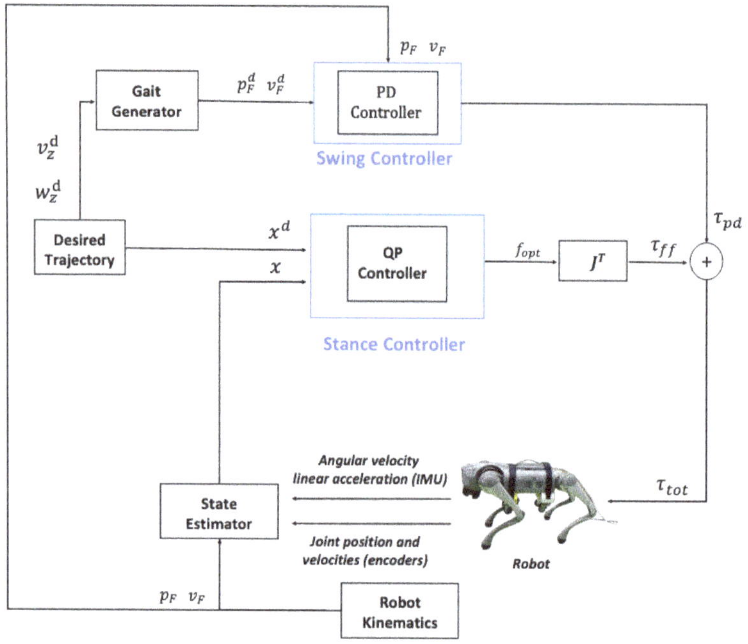

Figure 1. Control scheme.

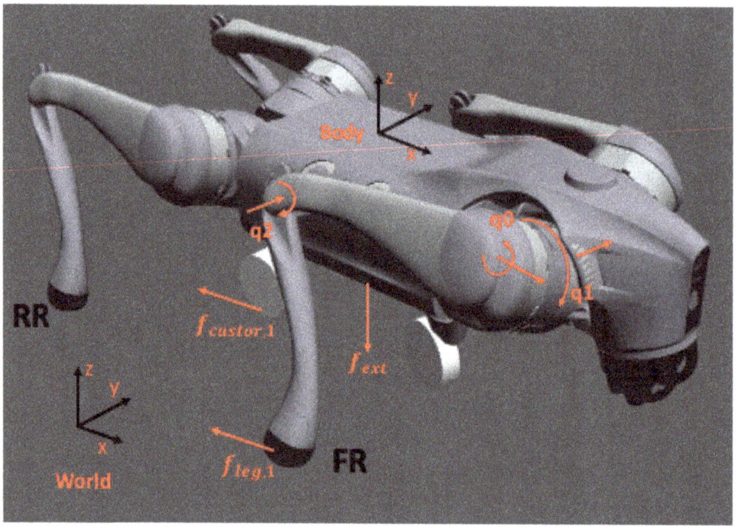

Figure 2. The simulated Go2 robot in sliding position. The joint angles are indicated for the FR leg, as are the world frame and the robot reference frame.

4. Simulation and Experimental Setup

Experiments were conducted both in simulation and on the physical quadrupedal robot (Unitree Go2). This dual approach made it possible to first validate the proposed sliding-based locomotion strategy under controlled and repeatable conditions in simulation and then confirm its effectiveness on real hardware.

Two gaits were analysed in this study: trotting and sliding. Each gait was tested at four target speeds (0.16, 0.22, 0.28, and 0.34 m/s), with four repetitions per condition, resulting in a total of 32 trials. Each trial covered a distance of 10 m, with the duration depending on the commanded speed.

Data were collected entirely onboard the robot by using its internal sensors and control system. The recorded variables included joint positions, velocities, torques, ground reaction forces, and control commands (feedforward, proportional, and derivative terms), as well as battery voltage and current, all sampled at 500 Hz.

Voltage and current were acquired from the robot's low-level state interface and logged together with the joint and IMU data. Specifically, the Unitree low-level interface provides the pair (V, I) through the fields `powerV` and `powerA` (accessible via `LowlevelState::getVA()`), which were logged at the same sampling rate as the other state variables. Electrical power was computed as $P(t) = V(t)\,I(t)$. The Cost of Transport (COT) was then derived accordingly. No additional external sensing hardware was employed; data acquisition relied on the robot's internal sensors and factory calibration, and run-to-run consistency was verified by repeating trials under identical conditions.

In the simulation environment, battery-related measurements such as voltage and current were not available; instead, current was estimated using the neural network model. Apart from this difference, the same logging structure was maintained wherever possible to ensure consistency between simulated and real datasets.

4.1. Simulation Framework

The simulation environment was created using Gazebo Classic and the DART (Dynamic Animation and Robotics Toolkit) physics engine [29], which was chosen for its precise handling of joint dynamics. Although DART does not allow for the fine-tuning of surface friction parameters in the Simulation Description Format (SDF), it provides solid joint-level physics, well-suited for modelling and controlling the robot's kinematic chain.

The ground–wheel friction coefficient was neither explicitly measured nor tuned in the simulator; the default value provided by the DART physics engine was used. Instead of tuning this parameter, the joint-level friction and damping coefficients were empirically adjusted to reproduce the smooth rolling and sliding behaviour observed in the real robot on a painted concrete surface. This approach ensured that the overall motion dynamics in simulation closely matched the experimental observations, even though the exact ground–wheel friction coefficient was not directly identified.

The simulated robot model was designed to closely resemble the real hardware. However, one notable discrepancy concerned the third joint: in simulation, it was modelled as a directly motorised joint, whereas in the real robot, it is actuated via a transmission system. To account for this difference, a correction factor of 1.4 was applied to the simulated torque of the third joint, which was adjusted experimentally by comparing the simulated and measured torque curves to improve the match with the real behaviour. This adjustment brought the simulated torque much closer to the real behaviour and was used consistently in all experiments with sliding and trotting gaits.

The passive ventral wheels were modelled as castor wheels with a vertical axis of rotation, using continuous joints to allow for free pivoting as in the real wheels. Physical properties such as mass, inertia, and shape were determined based on realistic estimates of the actual components.

The damping and friction of the joints were implemented as numerical coefficients on the joint level (they correspond to the viscous friction forces expressed in $\mathrm{N \cdot m \cdot s/rad}$). Their values (0.015 and 0.01, respectively) were empirically tuned to reproduce the behaviour of the real robot.

4.2. Real Robot Testing

The real experiments took place indoors on a smooth, painted concrete floor, which was very similar to the simulated flat terrain, as shown in Figure 3.

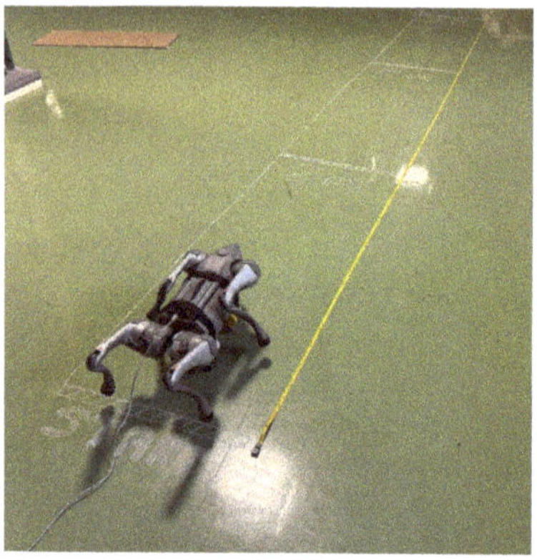

(a) Real environment

(b) Simulated environment

Figure 3. Visual comparison between the real terrain (a) and its simulated counterpart (b).

Chassis Modifications

To enable sliding locomotion, two removable support structures were mounted beneath the robot's belly. These supports held the castor wheels and were carefully designed to integrate seamlessly with the robot's existing walking capabilities. Their placement and geometry were planned so as not to interfere with leg motion or ground clearance.

The passive wheels were positioned symmetrically at the four corners of the main body, maintaining a balanced centre-of-mass distribution. Thanks to this configuration, the overall weight distribution remains stable during motion.

The additional ventral wheels slightly reduce the robot's ground clearance but do not compromise its ability to perform standard gaits such as trotting or crawling. The robot retains its full range of motion and can transition smoothly between walking and sliding phases, as shown in Figure 4.

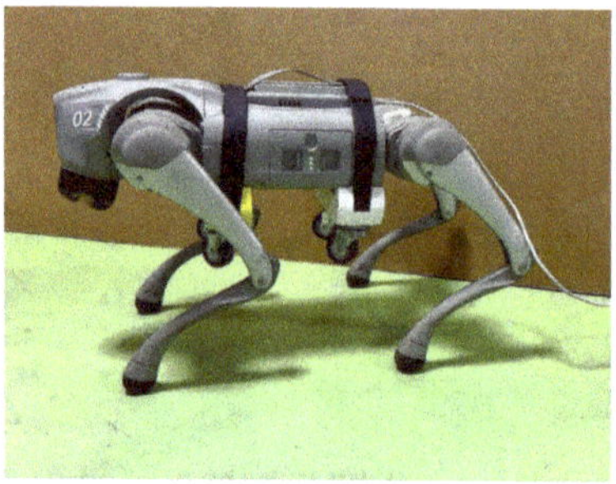

(a)

(b)

Figure 4. *Cont.*

(**c**) (**d**)

Figure 4. A sequence of four images showing the Unitree Go2 robot transitioning from a standing position (**a**) to a sliding configuration (**d**). The transformation occurs step by step through the intermediate stages (**b**,**c**).

The ventral wheels were designed to integrate seamlessly with the walking capabilities, without affecting ground clearance or leg motion. Commercial castor wheels were selected for their light weight, availability, and omnidirectional motion capability. The rubber coating provides sufficient grip while maintaining low rolling friction for smooth gliding. Each wheel was mounted using a 3D-printed support precisely shaped to fit the robot's torso, ensuring rigidity and symmetry across the four contact points while remaining easy to attach or remove when required. Further specifications of the ventral wheels are summarised in Table 3.

Table 3. Physical and geometric properties of the ventral wheels. Mounting height h corresponds to the total vertical distance from the ground to the robot torso during sliding, while abdomen clearance c represents the height of the supporting structure beneath the torso.

Property	Symbol / Unit	Value
Wheel type	–	Castor wheel
Material	–	Rubber
Diameter	D [mm]	50
Width	w [mm]	18
Mass (each)	m [g]	120
Bearing type	–	Ball-bearing swivel joint
Number of wheels	n	4
Torso height from ground (mounting height)	h [mm]	105
Support structure height (abdomen clearance)	c [mm]	51
Longitudinal spacing (front–rear)	L_{long} [mm]	200
Lateral spacing (left–right)	L_{lat} [mm]	130

4.3. Comparative Analysis

To check how well the simulation matched reality, we compared the torque and speed profiles of the joints. Figures 5 and 6 show typical torque curves for sliding and trotting gaits at a reference forward speed of 0.22 m/s, which prove the good accuracy of the simulation.

The availability of a simulation very close to the real behaviour of the robot is a first step that still requires further effort to estimate the power consumption in the simulated environment for COT analyses. For this purpose, a neural network was trained based on real data to predict the robot's power consumption, which enabled a consistent assessment of energy consumption in both the simulation and the real experiments.

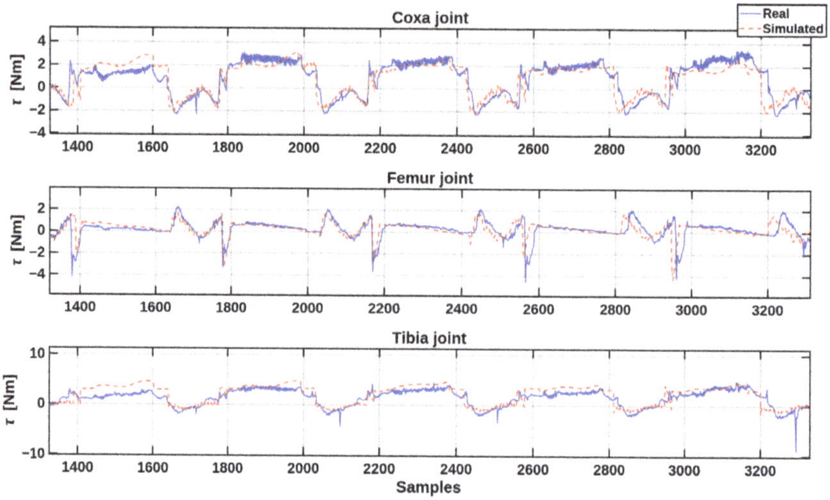

Figure 5. Comparison between real and simulated joint torques for one leg of the robot while sliding at 0.22 m/s.

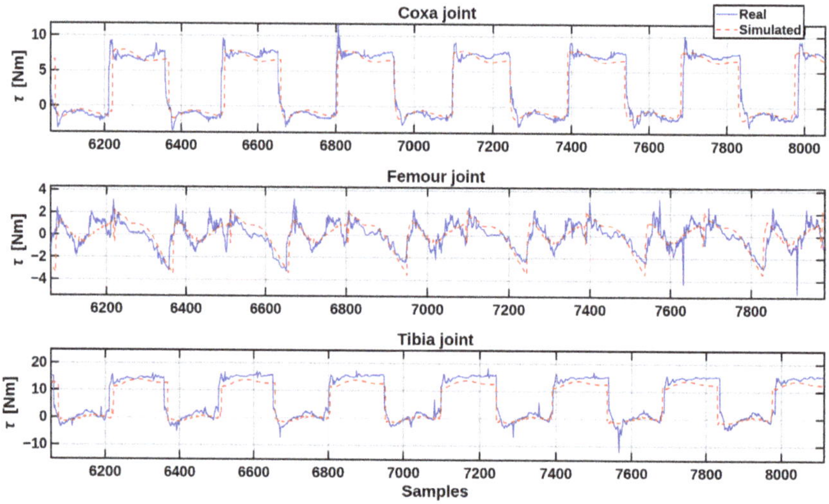

Figure 6. Comparison between real and simulated joint torques for one leg of the robot while trotting at 0.22 m/s.

Figure 7 shows representative foot trajectories in the sagittal plane (X-Z) calculated using forward kinematics from onboard joint measurements. Two gait patterns, sliding and trotting, are compared. While the sliding gait shows smoother and flatter trajectories due to the continuous support from the abdomen, the trotting gait shows more pronounced cycloid movement typical of the periodic lifting of the legs. This contrast reflects the different control settings and body postures. In the trotting gait, the robot performs shorter and faster steps (period $T = 0.6$ s, stance ratio = 0.5, leg tip height 0.08 m), resulting in higher, rounder trajectories that help maintain dynamic stability during the alternating foot support phases. The period T refers to the duration of the stride cycle, the stance ratio indicates the proportion of the cycle in which the foot is in contact with the ground, and the leg tip height corresponds to the vertical amplitude of the foot lift.

In contrast, the ventral wheels provide continuous support during sliding, which allows for longer and slower steps (period $T = 0.8$ s, stance ratio = 0.6, and leg tip height = 0.03 m), resulting in lower and longer foot travel. The lowered posture during sliding also increases the effective range of leg extension.

See also the supplementary video material (https://youtu.be/1UYCAia5hS8, (accessed on 22 September 2025)), which illustrates the foot trajectories during motion and provides a visual confirmation of the observed differences.

All values have been experimentally tuned to ensure stable and smooth locomotion in both gaits. To closely replicate the real behaviour of the robot, it became necessary to estimate power consumption within the simulated environment to compute the COT.

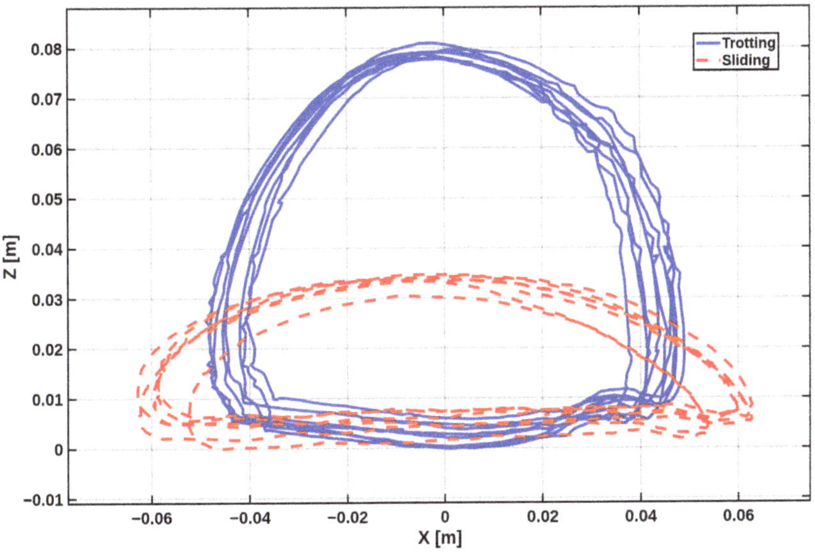

Figure 7. Comparison of representative foot trajectories in the sagittal plane (X–Z) for *trotting* (blue, solid) and *sliding* (red, dashed) gaits.

4.4. Data-Driven Current Estimation Model

To estimate the robot's power consumption during the simulation, we developed a neural network model that was trained exclusively on experimental data. The decision to use a neural network was driven by the complex and highly nonlinear nature of the robot's power consumption, which depends on joint torques, velocities, and their dynamic interactions. In addition, many parameters depend on hardware details (e.g. sensors, drivers, circuit boards, and onboard embedded computer) that are difficult to estimate analytically. However, these parameters play a crucial role in COT estimation. Traditional analytical or linear models proved insufficient to capture these features. In contrast, the neural network showed high estimation accuracy and generalisation under all tested conditions.

The neural model takes as input the joint torques and angular velocities of all 12 actuated joints, resulting in 24 input features, and produces a single scalar output representing the estimated total current absorbed by the robot. Figure 8 illustrates the network architecture.

The proposed neural network architecture consists of a fully connected feedforward multilayer perceptron with ReLU activation functions and the Mean Squared Error (MSE) loss function. The final output layer consists of a single linear neuron that produces the continuous prediction. A thorough grid search was performed over multiple hyperparameters, including the number of hidden layers, neurons per layer, learning rates, and batch sizes, to determine the optimal architecture and training configuration, as shown in Figure 9. The results show that both the network depth and the number of neurons per layer strongly influence the performance: models with at least four hidden layers and 32 or more neurons per layer consistently achieved higher R^2 values. The batch size also played a role, with medium-to-large values (256–512) giving the best results, while excessively large batch sizes degraded performance. In contrast, the learning rate mainly affected performance at extreme values, with most settings in the 3×10^{-5}–10^{-3} range achieving similarly good results. Therefore, the selected network configuration is characterised by four hidden layers, each comprising 256 neurons. Training was performed with the Adam optimiser using a learning rate of 3×10^{-4}, a batch size of 512, and early stopping based on validation loss.

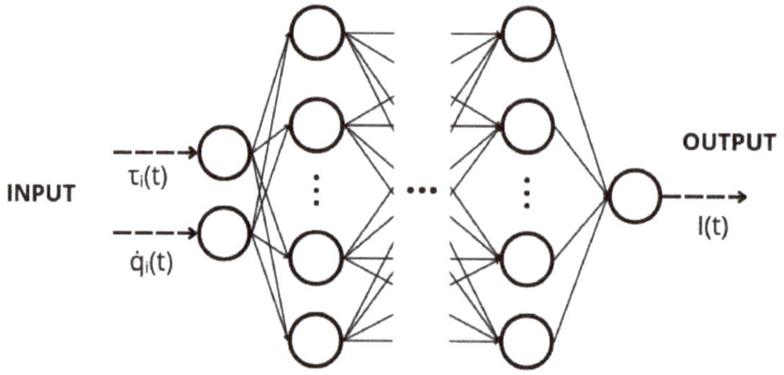

Figure 8. Schematic representation of the neural network for electrical current estimation.

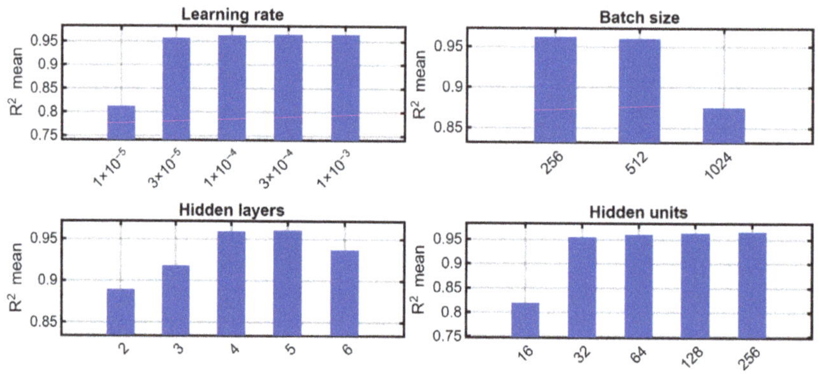

Figure 9. Impact of key hyperparameters on model performance (R^2). Bar plots report the average score across different configurations.

The dataset used for network training was built from on-board sensor measurements collected over multiple experimental trials. Specifically, 75% of the recorded runs—corresponding to three out of four repetitions for each gait and speed—were used to form the training and validation dataset, resulting in a total of 367,840 samples recorded at 500 Hz across all runs. Within this subset, the data were further divided into 80% for training and 20% for validation through random splitting. The remaining 25% of the runs (one repetition per condition) were kept completely separate and used as an independent test set to evaluate the model's generalisation capability. No data augmentation or input normalisation was applied, relying on the intrinsic consistency and bounded ranges of the experimental measurements.

The neural model achieves a Mean Squared Error (MSE) of 0.1674, a Mean Absolute Error (MAE) of 0.1684, and a coefficient of determination (R^2) of 0.9740 on the independent test set, indicating a strong generalisation ability. These results confirm that the model accurately reproduces the complex nonlinear relationship among joint dynamics, robot hardware, and current consumption. Additional error statistics, including the Mean Absolute Percentage Error (MAPE) and the 95th percentile of the absolute error, are reported in Table 4.

Table 4. Neural model performance on the independent test set.

Metric	Symbol	Value
Mean Squared Error	MSE	0.1674
Mean Absolute Error [A]	MAE	0.1684
Coefficient of determination	R^2	0.974
Mean Absolute Percentage Error [%]	MAPE	6.28
Std. deviation of abs. error [A]	$\sigma_{\lvert e \rvert}$	0.357
95th percentile of abs. error [A]	-	0.468

All metrics are computed on the independent test set using measured and predicted total current.

These results confirm that the proposed neural model effectively captures the nonlinear relationship between joint dynamics and electrical power consumption. Its low error and high R^2 make it suitable for use as a reliable data-driven estimator of the robot's instantaneous current, enabling accurate COT evaluation in simulation without requiring additional physical measurements.

Figure 10 shows a comparison between the measured current and the current predicted by the neural network over a period of 5 s. The prediction slightly underestimates the peak values but accurately reflects the overall temporal profile.

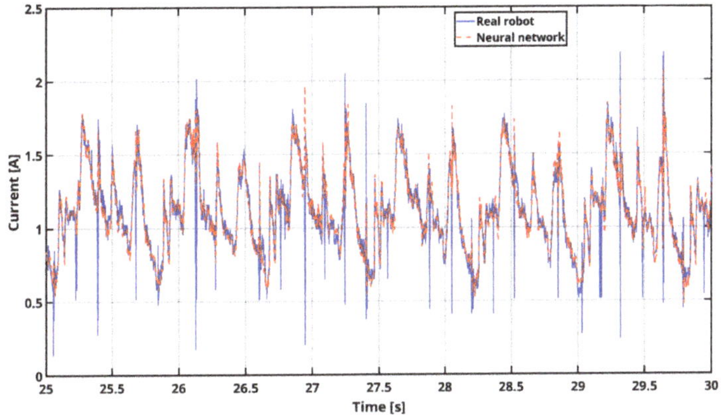

Figure 10. Comparison of the measured motor current (blue) and the current predicted by the neural network (dashed red) over a representative 5 s interval obtained during a sliding experiment at a constant speed of 0.22 m/s.

Figure 11 shows a regression diagram comparing the predicted and measured current values for the test set. The close correspondence of the data points with the diagonal line indicates that the model can accurately capture the magnitude and overall distribution of the real measurements.

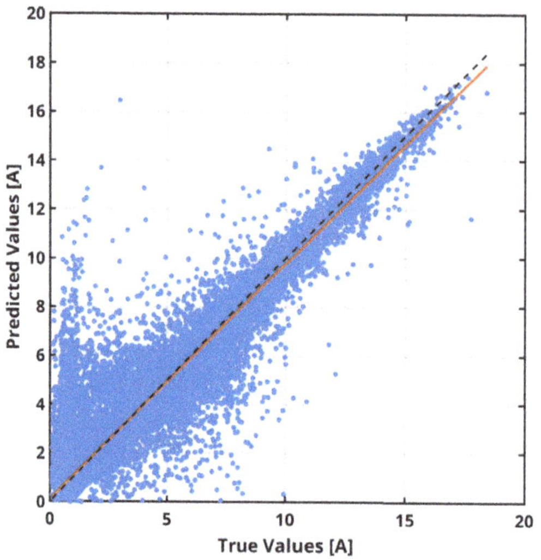

Figure 11. Regression plot showing the relation between measured and predicted current values over a test sequence. The 95% of the samples exhibit an absolute prediction error below 0.5 A (0.494 A).

4.5. Data Preparation

For each trial, the initial 1.5–2.0 s was discarded to remove start-up transients and compute all quantities under steady-state conditions. In the simulation, log files were parsed, and the torque of the third joint was scaled by a factor of 1.4 to match the real transmission, as detailed before. For the real-robot dataset, missing or null samples were flagged, and outliers were detected using a z-score threshold ($|z| > 4$) on each channel; flagged samples were then linearly interpolated along time (with nearest-neighbour fallback) to preserve continuity. The neural network inputs were constructed by concatenating joint torques and velocities (24 features), and the target was the electrical current, corresponding to the total current measured with the robot's power interface.

While the training and test data cover the nominal range of 0.16–0.34 m/s, the intrinsic variability and broadband nature of legged locomotion produce a wide spectrum of instantaneous velocities and dynamic conditions, which enrich the dataset with intrinsic noise within this interval. This enables the model to generalise reliably even beyond the trained speed range.

4.6. Sensitivity Analysis

To further interpret the neural model and assess the physical relevance of its predictions, the Pearson correlation coefficient [30] was computed between each input feature (joint torque or angular velocity) and both the measured and predicted total current. As shown in Table 5, the variable importance analysis indicates that joint torques are the primary drivers of current consumption, and that the hind legs are the main contributors during sliding, whereas the fore legs act predominantly for stabilisation. Measured and predicted correlation patterns match closely ($\Delta r < 0.01$), confirming that the model successfully captures the relevant physical dependencies.

Table 5. Top 10 input features ranked by absolute Pearson correlation with current.

Feature	Type	$\lvert r \rvert_{\text{real}}$	$\lvert r \rvert_{\text{pred}}$
T9	Torque	0.491	0.499
T7	Torque	0.456	0.464
T6	Torque	0.441	0.450
T8	Torque	0.426	0.428
T4	Torque	0.363	0.370
T12	Torque	0.357	0.354
T3	Torque	0.324	0.321
T10	Torque	0.307	0.303
T11	Torque	0.259	0.257
T1	Torque	0.232	0.228

Pearson correlation coefficients computed for each input feature with both the measured and predicted total current. The similarity between real and predicted patterns confirms that the neural model preserves the physical relevance of the joint torques.

4.7. Cost of Transport

The main motivation for estimating the motor current was to enable an accurate calculation of the robot's COT [8], which is defined as follows:

$$\text{COT} = \frac{P_{\text{E}}}{mgv} \tag{14}$$

where P_{E} is the total electrical power consumption, m is the robot's mass, g is the gravitational acceleration, and v is the average forward velocity. Electrical power was computed as

$$P_{\text{E}} = V_b \cdot I \tag{15}$$

where I is the electrical current and V_b the battery voltage. On the real robot, both I and V_b were directly measured with onboard sensors. In the simulation, current was estimated through the neural network, while the battery voltage was assumed constant at the nominal voltage of the real system (28.8 V). This assumption was verified through experimental measurements: during real trials, the onboard voltage remained highly stable around its nominal value, with a mean variation of about 0.7 V ($\pm$1.0 V standard deviation) and intra-trial fluctuations below 1.3%. Since electrical power is computed as $P_{\text{E}} = V_b I$, a voltage change of $\pm$1 V ($\approx$3.5%) would induce a comparable relative variation in the estimated COT. Such an effect is negligible compared with the observed $\approx$80% difference between gaits, confirming that the constant-voltage approximation introduces only a minor error.

This approach allowed us to compute a realistic and consistent estimate of P_{E}, and consequently of the COT, in both simulations and real experiments.

The Cost of Transport (COT) was computed for both trotting and sliding gaits based on measured and NN-estimated power consumption across different commanded speeds. Figure 12 illustrates the distribution of COT values, while Table 6 summarises the corresponding numerical results (mean $\pm$ std).

As shown in Table 6, the neural model reproduces the experimental trends with remarkable consistency, both across gaits and speeds.

This close correspondence enables straightforward calculation of the power consumption and thus the COT in the simulation. This careful validation was crucial to ensuring that our neural network for estimating motor current works reliably under both simulated and real-world conditions.

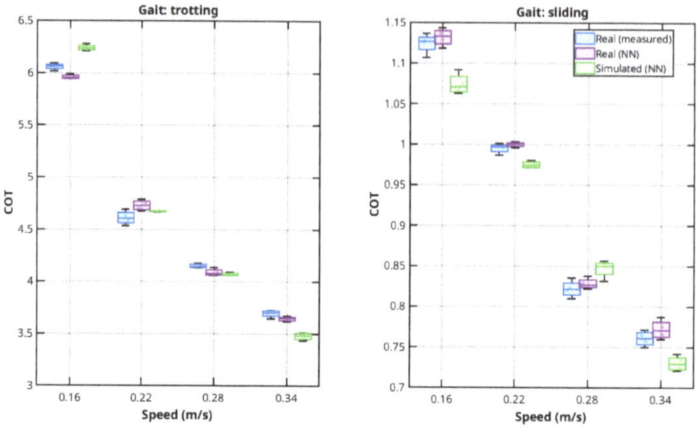

Figure 12. Box plots of the COT for different gaits and speeds. The results are shown for two locomotion gaits (trotting and sliding). For each speed, three datasets are compared: experimental measurements on real trials ("Real"), predictions of the neural network on the same real data ("Real NN"), and predictions of the neural network on simulated data ("Sim NN").

Table 6. Cost of Transport (mean ± std) for trotting and sliding at four commanded speeds.

Gait	Source	0.16 m/s	0.22 m/s	0.28 m/s	0.34 m/s
Trotting	Real (measured)	6.06 ± 0.03	4.61 ± 0.07	4.15 ± 0.02	3.69 ± 0.03
	Real (NN)	5.96 ± 0.02	4.73 ± 0.05	4.08 ± 0.03	3.64 ± 0.02
	Sim (NN)	6.24 ± 0.03	4.67 ± 0.01	4.07 ± 0.02	3.48 ± 0.04
Sliding	Real (measured)	1.12 ± 0.01	1.00 ± 0.01	0.82 ± 0.01	0.76 ± 0.01
	Real (NN)	1.13 ± 0.01	1.00 ± 0.00	0.83 ± 0.01	0.77 ± 0.01
	Sim (NN)	1.07 ± 0.01	0.97 ± 0.00	0.85 ± 0.01	0.73 ± 0.01

The present study already demonstrates that the sliding gait achieves a COT approximately five times lower than that of the trotting gait. To formally assess the significance and consistency of this method, a statistical analysis was performed as described in the following section. Although the addition of wheels slightly increases the total weight of the robot, the corresponding COT remains very close to that of the standard configuration. This is mainly due to the relatively small increase in mass (around 4%), which only has a marginal effect on the overall energy consumption. From Equation (14), it results that the impact of this added mass on the normalised COT is negligible: any small increase in mass produces an equivalent effect on the required power, leading to nearly unchanged COT values.

In addition, the robot exhibits greater stability in the sliding configuration, as it has four points of contact (the passive belly wheels) with the ground. This improved stability, combined with the significantly reduced COT values, makes the sliding gait a promising locomotion strategy on flat or slightly inclined surfaces. Overall, the experiments confirm that the robot maintains an optimal posture and achieves speeds comparable to those of a trot while benefiting from the improved stability and lower energy costs of sliding.

Statistical Validation

Two complementary analyses were conducted: Welch's unequal-variance *t*-tests [31], to compare mean COT values between gaits for each speed and data source, and a two-way ANOVA [32], to assess the influence of gait and speed, as well as their interaction.

Welch's *t*-tests confirmed that the mean COT of the sliding gait is significantly lower than that of the trotting gait across all speeds and data sources ($p < 0.001$), with an average reduction of about 80%. This demonstrates that the observed energy savings are consistent and not due to random variability, as also evidenced by Table 7.

Table 7. Welch's t-test results comparing trotting and sliding gaits. Values are reported as mean ± standard deviation. All differences are statistically significant ($p < 0.001$).

Source	Speed [m/s]	Trot	Slide	Ratio (T/S)	Δ [%]
Real (measured)	0.16	6.06 ± 0.03	1.12 ± 0.01	5.4	81.4
	0.22	4.61 ± 0.07	1.00 ± 0.01	4.6	78.4
	0.28	4.15 ± 0.02	0.82 ± 0.01	5.0	80.2
	0.34	3.69 ± 0.03	0.76 ± 0.01	4.9	79.4
Real (NN)	0.16	5.96 ± 0.02	1.13 ± 0.01	5.3	81.0
	0.22	4.73 ± 0.05	1.00 ± 0.00	4.7	78.9
	0.28	4.08 ± 0.03	0.83 ± 0.01	4.9	79.7
	0.34	3.64 ± 0.02	0.77 ± 0.01	4.7	78.8
Sim (NN)	0.16	6.24 ± 0.03	1.07 ± 0.01	5.8	82.8
	0.22	4.67 ± 0.01	0.97 ± 0.00	4.8	79.1
	0.28	4.07 ± 0.02	0.85 ± 0.01	4.8	79.2
	0.34	3.48 ± 0.04	0.73 ± 0.01	4.8	79.0

Δ: Mean percentage reduction in COT for sliding relative to trotting gaits. All pairwise Welch's *t*-tests yielded $p < 0.001$ and large effect sizes ($\eta^2 > 0.99$), confirming that the energy reduction of approximately 80% is highly significant across all conditions.

The two-way ANOVA (Gait × Speed) confirmed that both gait and commanded speed have statistically significant effects on the COT, as well as a strong interaction between the two factors (Table 8). This analysis demonstrates that the energy efficiency advantage of sliding remains robust across all tested speeds and datasets.

Table 8. Two-way ANOVA (Gait × Speed) results for each data source.

Source	*p* (Gait)	*p* (Speed)	*p* (Interaction)
Real (measured)	5.98×10^{-46}	3.45×10^{-31}	5.78×10^{-28}
Real (NN)	9.84×10^{-48}	6.42×10^{-33}	1.43×10^{-29}
Sim (NN)	4.66×10^{-51}	1.68×10^{-37}	5.43×10^{-35}

Both gait and speed have highly significant effects ($p < 10^{-30}$), and their interaction confirms that the reduction in COT with sliding is consistent but slightly modulated by velocity.

Overall, these results demonstrate that the sliding gait reduces energy expenditure by approximately 80% while maintaining stable and repeatable performance across speeds and datasets. This confirms the validity of the proposed estimation framework and its suitability for guiding future data-driven gait optimisation.

5. Conclusions

In this work, we presented a novel hybrid locomotion strategy for quadrupedal robots that combines traditional legged gait with sliding assisted by passive wheels mounted on the abdomen. In addition to this mechanical innovation, we have introduced a neural network model that can estimate the motor current directly from the joint dynamics, which enables a consistent calculation of the COT in the simulation.

This reverse-engineering approach allows for extensive use of the simulation environment by compensating for missing information through reliable estimations derived from the real robot. The main advantage is the ability to rely on simulation for testing and validation, thereby reducing mechanical stress on the hardware and limiting the need for time-consuming experimental campaigns.

This limitation in friction control was considered an acceptable trade-off between surface friction and accurate joint dynamics. In our case, the main goal was to reproduce realistic actuator behaviour and controller response, which DART handles reliably. Even with simplified friction modelling, the simulated sliding motion showed good agreement with the real robot, both in its kinematic behaviour and in overall energy trends. If future studies require simulating terrains with very different friction characteristics (for example, ice or soft soil), other physics engines could be used to achieve finer control of surface properties, although this would likely reduce the accuracy of the joint dynamics. Overall, the current setup proved effective: the real and simulated controllers performed consistently, and any small differences could be corrected by tuning the controller parameters.

The results clearly show that the sliding gait achieves significant energy savings, with a COT that is almost five times lower than that of the classic trot. This is particularly remarkable considering that the robot maintains a speed and stability comparable to those of a trot thanks to the continuous ground contact of the ventral wheels.

A neural network was developed to estimate the current consumption for different gaits and speeds. This enables reliable estimation of COT in simulation and paves the way for safer and more realistic gait optimisation before hardware implementation.

It is important to note that the effectiveness of the sliding gait is limited to flat or slightly descending terrain due to the relatively small radius of the wheels and their reduced ability on uneven surfaces. On uneven or obstacle-strewn terrain, the trot is still the preferred gait. However, in indoor or industrial environments, sliding is a very favourable solution to reduce energy consumption without sacrificing performance.

This result is also in line with the biologically inspired foundations of our work. In developing the hybrid gait, we were inspired by animals such as penguins, otters, and turtles, which utilise sliding on low-friction surfaces to reduce muscle effort and improve energy efficiency. Similarly, our robot uses passive ventral wheels to alternate between active locomotion and energy-saving sliding phases. This parallel between biological and robotic locomotion emphasises the effectiveness of sliding as a natural strategy to achieve efficiency and stability on suitable terrain.

This type of locomotion could be particularly useful in semi-structured environments such as industrial plants or tunnels, as these environments are typically characterised by the presence of flat paths, as well as ladder or single stairs. Here, the advantages of our approach are clear; of course, to maximise efficiency, a parallel algorithm able to classify the type of terrain and autonomously switch between sliding and other legged locomotion patterns has to be conveniently used. Regarding this point, in the literature, several efficient solutions are already available.

The sliding conditions, including friction cones, contact constraints, and passive wheel handling, are currently enforced via a QP. In future work, these constraints could be integrated into a reinforcement learning framework by embedding them into the reward function, penalising violations and rewarding compliant actions. Additionally, incorporating the robot's total electrical power into the reward would promote energy-efficient locomotion, encouraging smoother motions and minimising unnecessary actuator effort while maintaining physically consistent sliding behaviour. Such an approach could further enhance efficiency and adaptability, enabling the policy to optimise gait selection based on terrain and payload, to maximise energetic performance.

In addition, the dynamic applicability of the proposed sliding approach depends on the terrain slope and the robot's speed. While mild downhill surfaces can further reduce the COT thanks to gravity assistance, steeper slopes would require active braking or specific control strategies to keep the robot stable and manoeuvrable. Therefore, although the sliding gait remains highly efficient under favourable conditions, safe operation in more dynamic scenarios will depend on the implementation of dedicated control mechanisms.

Author Contributions: Conceptualisation, methodology, software, investigation, and writing: D.O.A.T., A.L.N., L.P., and P.A.; writing—review and editing: D.O.A.T., A.L.N., L.P., and P.A.; Supervision and project administration: L.P. and P.A. All authors have read and agreed to the published version of the manuscript.

Funding: The work was supported by MUR PNRR—Mission 4-Comp.2—Inv:1.3—project PE0000013-FAIR.

Data Availability Statement: The data presented in this study are available on request from the corresponding author. The data are not publicly available due to their large size and internal storage format.

Conflicts of Interest: The authors declare no conflicts of interest.

References

1. Iida, F.; Ijspeert, A.J. Biologically Inspired Robotics. In *Springer Handbook of Robotics*; Siciliano, B., Khatib, O., Eds.; Springer: Cham, Switzerland, 2016; pp. 2015–2034. [CrossRef]
2. Hoeller, D.; Rudin, N.; Sako, D.; Hutter, M. ANYmal parkour: Learning agile navigation for quadrupedal robots. *Sci. Robot.* **2024**, *9*, eadi7566. [CrossRef] [PubMed]
3. Kim, S.; Wensing, P. Design of Dynamic Legged Robots. *Found. Trends Robot.* **2017**, *5*, 117–190. [CrossRef]
4. Katz, B.; Carlo, J.D.; Kim, S. Mini Cheetah: A Platform for Pushing the Limits of Dynamic Quadruped Control. In Proceedings of the Proceedings of the 2019 International Conference on Robotics and Automation (ICRA), Montreal, QC, Canada, 20–24 May 2019; pp. 6295–6301. [CrossRef]
5. Boston Dynamics. Spot Robot. 2025. Available online: https://bostondynamics.com/products/spot/ (accessed on 15 September 2025).
6. Unitree Robotics. Unitree Robotics—Go2. 2025. Available online: https://www.unitree.com/go2 (accessed on 15 September 2025).
7. Arena, P.; Patanè, L.; Taffara, S. Energy Efficiency of a Quadruped Robot with Neuro-Inspired Control in Complex Environments. *Energies* **2021**, *14*, 433. [CrossRef]
8. Lei, J.; Wang, F.; Yu, H.; Wang, T.; Yuan, P. Energy Efficiency Analysis of Quadruped Robot with Trot Gait and Combined Cycloid Foot Trajectory. *Chin. J. Mech. Eng.* **2014**, *27*, 138–145. [CrossRef]
9. Silva, M.F.; Machado, J.A.T. Energy Efficiency of Quadruped Gaits. In *Climbing and Walking Robots*; Tokhi, M.O., Virk, G.S., Hossain, M.A., Eds.; Springer: Berlin/Heidelberg, Germany, 2006; pp. 735–742. [CrossRef]
10. Zhu, W.; Rosendo, A. PSTO: Learning Energy-Efficient Locomotion for Quadruped Robots. *Machines* **2022**, *10*, 185. [CrossRef]
11. Hutter, M.; Gehring, C.; Jud, D.; Lauber, A.; Bellicoso, C.D.; Tsounis, V.; Hwangbo, J.; Bodie, K.; Fankhauser, P.; Bloesch, M.; et al. ANYmal—A Highly Mobile and Dynamic Quadrupedal Robot. In Proceedings of the 2016 IEEE/RSJ International Conference on Intelligent Robots and Systems (IROS), Daejeon, Republic of Korea, 9–14 October 2016; pp. 38–44. [CrossRef]
12. Bjelonic, M.; Bellicoso, C.D.; de Viragh, Y.; Sako, D.; Tresoldi, F.D.; Jenelten, F.; Hutter, M. Keep Rollin'—Whole-Body Motion Control and Planning for Wheeled Quadrupedal Robots. *IEEE Robot. Autom. Lett.* **2019**, *4*, 2116–2123. [CrossRef]
13. Kashiri, N.; Baccelliere, L.; Muratore, L.; Laurenzi, A.; Ren, Z.; Hoffman, E.M.; Kamedula, M.; Rigano, G.F.; Malzahn, J.; Cordasco, S.; et al. CENTAURO: A Hybrid Locomotion and High Power Resilient Manipulation Platform. *IEEE Robot. Autom. Lett.* **2019**, *4*, 1595–1602. [CrossRef]
14. Zhou, Q.; Yang, S.; Jiang, X.; Zhang, D.; Chi, W.; Chen, K.; Zhang, S.; Li, J.; Zhang, J.; Wang, R.; et al. Max: A wheeled-legged quadruped robot for multimodal agile locomotion. *IEEE Trans. Autom. Sci. Eng.* **2024**, *21*, 7562–7582. [CrossRef]
15. Pan, Y.; Khan, R.A.I.; Zhang, C.; Zhang, A.; Shang, H. Pegasus: A Novel Bio-inspired Quadruped Robot with Underactuated Wheeled-Legged Mechanism. In Proceedings of the Proceedings of the 2024 IEEE International Conference on Robotics and Automation (ICRA), Yokohama, Japan, 13–17 May 2024; pp. 1464–1469. [CrossRef]
16. Magdaleno, E.C.O.; Cafolla, D.; Ceccarelli, M.; Castañeda, E.C.; Carbone, G. Experiences for a User-Friendly Operation of Cassino Hexapod III. In *Advances in Service and Industrial Robotics. RAAD 2018. Mechanisms and Machine Science*; Aspragathos, N., Koustoumpardis, P., Moulianitis, V., Eds.; Springer: Cham, Switzerland, 2019; Volume 67. [CrossRef]
17. Žák, M.; Rozman, J.; Zbořil, F.V. Energy Efficiency of a Wheeled Bio-Inspired Hexapod Walking Robot in Sloping Terrain. *Robotics* **2023**, *12*, 42. [CrossRef]
18. NASA Jet Propulsion Laboratory. RoboSimian. 2015. Available online: https://www.jpl.nasa.gov/robotics-at-jpl/robosimian/ (accessed on 15 September 2025).

19. Chen, J.; Xu, K.; Qin, R.; Ding, X. Locomotion control of quadrupedal robot with passive wheels based on CoI dynamics on SE(3). *IEEE Trans. Ind. Electron.* **2024**, *71*, 7551–7560. [CrossRef]
20. Singh, H. Future to Wheeled and Legged Robot Locomotion. *Int. J. Multidiscip. Res. (IJFMR)* **2025**, *7*, IJFMR250337131.
21. Chen, J.; Qin, R.; Huang, L.; He, Z.; Xu, K.; Ding, X. Unlocking Versatile Locomotion: A Novel Quadrupedal Robot with 4-DoFs Legs for Roller Skating. In Proceedings of the Proceedings of the 2024 IEEE International Conference on Robotics and Automation (ICRA), Yokohama, Japan, 13–17 May 2024; pp. 8037–8043. [CrossRef]
22. Davenport, J.; Bels, V. Quadrupedal terrestrial locomotion in emperor penguins (*Aptenodytes Forsteri*). *J. Nat. Hist.* **2023**, *57*, 1972–1983. [CrossRef]
23. Stevens, S.; Serfass, T. Sliding behavior in Nearctic river otters: Locomotion or play? *Northeast. Nat.* **2005**, *12*, 241–244. [CrossRef]
24. Godon, S.; Kruusmaa, M.; Ristolainen, A. Maneuvering on non-Newtonian fluidic terrain: A survey of animal and bio-inspired robot locomotion techniques on soft yielding grounds. *Front. Robot. AI* **2023**, *10*, 1113881. [CrossRef] [PubMed]
25. Haug, E. Simulation of friction and stiction in multibody dynamics model problems. *Mech. Based Des. Struct. Mach.* **2018**, *46*, 296–317. [CrossRef]
26. Boyd, S.; Vandenberghe, L. *Convex Optimization*; Cambridge University Press: Cambridge, UK, 2004; Chapter 4.4.1.
27. Nocedal, J.; Wright, S.J. *Numerical Optimization*, 2nd ed.; Springer: New York, NY, USA, 2006; Chapter 16.
28. Li Noce, A.; Patanè, L.; Arena, P. A stable and safe method for two-leg balancing of a quadruped robot using a neural-network-based controller. *Robot. Auton. Syst.* **2025**, *186*, 104901. [CrossRef]
29. Lee, J.; Grey, M.X.; Ha, S.; Kunz, T.; Jain, S.; Ye, Y.; Srinivasa, S.; Stilman, M.; Liu, C.K. DART: Dynamic Animation and Robotics Toolkit. *J. Open Source Softw.* **2018**, *3*, 500. [CrossRef]
30. Benesty, J.; Chen, J.; Huang, Y.; Cohen, I. Pearson correlation coefficient. In *Noise Reduction in Speech Processing*; Springer: Berlin/Heidelberg, Germany, 2009; pp. 1–4. [CrossRef]
31. Welch, B.L. The generalization of Student's problem when several different population variances are involved. *Biometrika* **1947**, *34*, 28–35. [CrossRef] [PubMed]
32. Fisher, R.A. *Statistical Methods for Research Workers*; Oliver and Boyd: Edinburgh, UK, 1925.

robotics

MDPI

Article

Real-Time Forecasting of a Fire-Extinguishing Agent Jet Trajectory from a Robotic Fire Monitor Under Disturbances

Irina Pozharkova * and Sergey Chentsov

Department of Automation Systems, Automated Control and Design, Siberian Federal University, 660041 Krasnoyarsk, Russia; schentsov@sfu-kras.ru
* Correspondence: ipozharkova@sfu-kras.ru

Abstract

This article presents a methodology for real-time forecasting of a fire-extinguishing agent jet trajectory from a robotic fire monitor under wind influence, which can significantly displace the impact area position and complicate targeting. The proposed methodology is designed for controlling firefighting robots in conditions where visual monitoring of the impact area is impeded by factors such as: obscuration of the fire-extinguishing agent flow by smoke, low visibility of its fragmented particles against the background environment, and long-range jet discharge. Trajectory forecasting is implemented using a neural network model. The training and verification of this model are performed with datasets constructed from the results of numerical simulations of fire-extinguishing agent motion under wind influence, based on Computational Fluid Dynamics (CFD) methods. Experimentally obtained data are used for the validation of the trained neural network model and the selected CFD models. The paper describes the methodology for conducting full-scale tests of fire monitors; a photogrammetric algorithm for generating validation datasets from the test results; an algorithm for calculating target characteristics, which describe the jet trajectory and are consistent with experimental data, used for forming training and verification datasets based on simulation; and a procedure for selecting Computational Fluid Dynamics models and their parameters to ensure the required accuracy. The article also presents the results of an experimental evaluation of the developed methodology's effectiveness for real-time prediction of the water jet trajectory from a fire monitor under various control and disturbance parameters.

Keywords: robotic fire monitor; neural network model; jet trajectory; real-time forecasting; computational fluid dynamics (CFD); photogrammetry

1. Introduction

Firefighting robots currently represent one of the most effective means of fire protection, enabling the suppression of fires with water jets or other extinguishing agents without direct human presence in zones exposed to hazardous fire factors [1]. At certain facilities, the task of ensuring fire safety is often complicated by the risk of collapse of load-bearing structures or tanks containing flammable substances due to heating to critical temperatures. In such cases, firefighting robots are also employed for cooling various protected objects [2]. Analysis of current research [1,3] indicates that the development of firefighting robots is focused on the creation of autonomous or teleoperated mobile platforms [4], including those based on unmanned aerial vehicles (UAVs) [5], enhancement of fire-seat detection algorithms [6], improvement of targeting and guidance systems for directing extinguishing agent jets using artificial intelligence and computer vision [7,8], and increasing the reliability of mechanical systems, among other areas.

A key control objective for firefighting robots is directing the flow of fire-extinguishing agent (FEA) towards a fire source, an object requiring cooling, etc. As shown in [9], achieving this in open spaces can be significantly complicated by turbulence and external factors. Specifically, wind can alter the jet trajectory and cause its disintegration, while low temperatures reduce jet height and range by altering the physical parameters of the ambient air and the extinguishing agent. This can ultimately lead to a significant displacement of the impact area (Figure 1).

Figure 1. Displacement of a water jet's impact area from a fire monitor under crosswind influence.

Turbulence significantly influences the solution of fire suppression jet targeting and control problems. The flow at the nozzle outlet of a fire monitor has a high velocity, which induces the formation of vortices at its boundaries. This leads to the mixing of the extinguishing agent with the ambient air through which it travels, resulting in the gradual breakup of the jet into fragments (disintegration) and individual droplets (atomization), an increase in its cross-sectional area, and a deceleration of its core [10]. The disintegrated and atomized portions of the extinguishing agent flow are substantially more susceptible to external disturbances, particularly wind, compared to the compact (coherent) part of the jet. As can be seen in Figure 1, the deviation of the jet in its initial region is practically negligible, whereas after its disintegration, the displacement of the impingement zone reaches a significant magnitude. The breakup of the FEA flow and its turbulent mixing with air complicate the targeting of the jet and can reduce the effectiveness of fire suppression or structural cooling due to a substantial decrease in the application density (the discharge rate of the extinguishing agent per unit area of the protected surface) [11].

Under such conditions, solving the targeting problem using visual feedback can be hindered by the following factors: smoke obscuration, visual noise from external objects overlapping the agent flow, and poor distinguishability of its fragmented part against a background of similar brightness and color. When targeting with a compact (solid) segment of the jet or a short-range jet, one can use trajectory data obtained from field tests under various control parameters without accounting for wind, as the influence of disturbances is less pronounced in this case. However, this approach may be ineffective for controlling a long-range, elevated trajectory jet. Conducting field tests to obtain empirical data describing the behavior of a fire robot's jet across a wide spectrum of external factor parameters, corresponding to various usage scenarios, is extremely challenging. Furthermore, the trajectory of the fire-extinguishing agent flow under wind influence represents a geometrically complex figure of potentially significant dimensions.

Photogrammetry [12] is one of the most accessible methods for 3D reconstruction of real objects. Currently, established methodologies and their software implementations exist [13], allowing for highly accurate creation of 3D models of various objects, buildings, structures, etc., based on multiple photographs taken from different viewpoints, including those from unmanned aerial vehicles (UAVs) [14]. However, the applicability of these methods is largely limited to static objects, whose shape and position remain stable during the "scanning" process. Consequently, standard photogrammetry algorithms may be inefficient for reconstructing 3D models of free jets, whose boundaries change substantially during motion due to various factors, particularly turbulence. Furthermore, this task is complicated by the following characteristics of the studied flow processes:

- Varying optical density of the jet along its length.
- Dependence of the optical density of the same jet fragment on the camera viewpoint.
- Poor distinguishability of the jet against a background of similar brightness and color (e.g., a water flow from a fire monitor against a cloudy sky).
- Visual noise of the semi-transparent jet by external objects in the background.
- Limited jet lifetime (e.g., during field tests without access to a continuous supply of the test fluid).
- Significant jet dimensions (e.g., tens of meters during fire monitor field tests).

These characteristics can complicate tasks such as obtaining synchronized, high-quality images of the process from a sufficient number of viewpoints, boundary extraction, feature point detection [15], image rectification [16], etc. Difficulties related to optical density, visual noise, and poor distinguishability are particularly pronounced in the jet breakup and spray regions, which are often of primary interest for targeting control.

In [17], a method for jet trajectory determination based on UAV imagery using trainable prompt vectors is presented. This method demonstrated high efficiency in tracking the fire-extinguishing agent flow without accounting for wind and potential image noise from smoke in fire conditions. In [18,19], a method for predicting jet trajectory based on a short-range computer vision system is described. This approach can be used in smoky conditions; however, under strong crosswinds and at long ranges, predicting the impact area position based on the jet's compact core may yield low accuracy. This is because the solid flow segment deviates only slightly from its initial direction, so an estimate of trajectory change over a long distance may differ significantly from the actual displacement. In [20], a method for predicting water jet falling point position under wind influence, based on a neural network model, was proposed and showed relatively high accuracy. However, that study used a water cannon with a short range, and the falling point was predicted only in a single horizontal plane. To enable jet targeting to any point within the fire robot's operational radius, information about the position of the extinguishing agent flow along its entire trajectory is required.

In [9], a method for predicting fire-extinguishing agent jet trajectory under wind load, based on Computational Fluid Dynamics (CFD) methods, was proposed, along with experience in using this method accounting for the influence of sub-zero temperatures. The primary drawback of this approach in the context of fire robot control is its significant computational complexity, which precludes its use in real-time mode, where simulation must proceed at the speed of the actual physical processes. To address this problem, various approaches are currently employed, including the following:

- Model simplification by increasing computational grid cell size [21], which can significantly reduce simulation time at the cost of calculation accuracy. This approach is largely ineffective for predicting fire-extinguishing agent trajectory, as it is difficult to achieve both acceptable error and prediction speed corresponding to real-time operation.
- Increasing the speed of numerical solution of differential equations by reducing the number of iterations based on the Arnoldi method [22]. This typically results in a minor

increase in error. However, for predicting fire-extinguishing agent trajectory under disturbances, this approach does not achieve the speed required for real-time operation.

- Reducing simulation time via parallel computing for solving CFD equations [23] while maintaining high accuracy. However, the application of this approach to fire robots is significantly constrained by the available hardware, which cannot provide acceptable calculation speeds.
- Calculating specific characteristics of the studied processes, such as impact area position, using machine learning [20]. This approach can potentially provide the required computational speed. However, constructing sufficient and high-quality training and validation datasets for configuring such models is a complex problem, as fire-extinguishing agent jets have significant geometric dimensions, dynamically changing boundaries, variable optical density, etc. These characteristics significantly complicate the measurement of the studied parameters via field tests.

The hypothesis of this work is that the combined use of machine learning (ML) and Computational Fluid Dynamics [24,25] can solve this problem. In this approach, numerical simulation of gas–liquid flow processes is used to generate datasets across a wide range of control and disturbance parameters. The data thus obtained are then used to train and verify neural network models intended for real-time estimation of relevant jet parameters. Validation datasets are formed based on field tests of robotic firefighting systems or their actuators (fire monitors). Implementing this approach is only possible if the target characteristics describing the dynamics of the studied flow processes are consistent between simulation and experimental results.

In this study, the primary target characteristic of the fire-extinguishing agent jet from a fire monitor is its trajectory, described by the upper boundary of the studied gas–liquid flow. This choice is motivated by the fact that it enables the determination of instrumentally measurable jet parameters: range, height, and coordinates of the farthest point of the impact area. These characteristics can be used for experimental validation of the obtained results.

To implement the proposed approach, the following tasks were accomplished within this study:

1. Development of a methodology for real-time prediction of fire-extinguishing agent flow trajectory.
2. Development of a methodology for reconstructing the fire-extinguishing agent flow trajectory from experimental data using photogrammetry methods for forming validation datasets.
3. Development of a methodology for calculating the fire-extinguishing agent flow trajectory from simulation results that is consistent with experimental data.
4. Selection of Computational Fluid Dynamics methods for simulating the motion of the fire-extinguishing agent under disturbances.
5. Experimental evaluation of the effectiveness of the developed methodology for real-time water jet trajectory prediction.

This article is structured as follows: Section 2 presents the methods used to solve the aforementioned tasks; Section 3 presents the results of reconstructing the fire-extinguishing agent flow trajectory from experiments, the selection of CFD methods for simulating the agent's motion under disturbances, and the evaluation of the developed real-time water jet trajectory prediction methodology's effectiveness; Section 4 presents conclusions and directions for future research.

2. Materials and Methods

2.1. Methodology for Real-Time Forecasting of Fire-Extinguishing Agent Trajectory

The core of the developed methodology is a trained computational model (Figure 2).

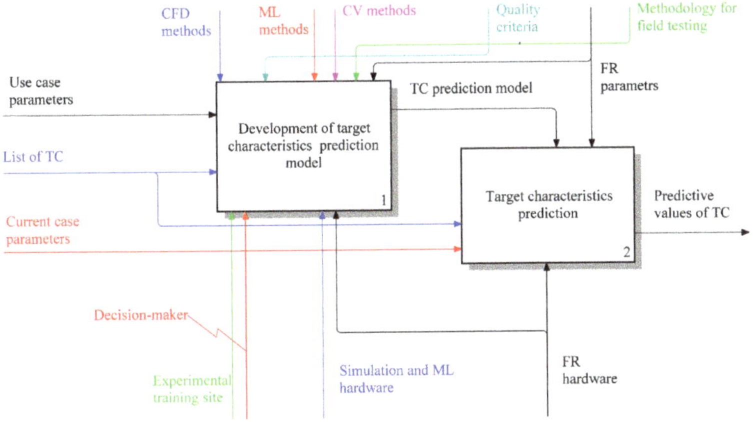

Figure 2. IDEF0 functional diagram for real-time forecasting of fire-extinguishing agent jet trajectory.

The construction of this model (Figure 2, block 1) consists of four main stages (Figure 3): initial data generation, machine learning, verification, and validation.

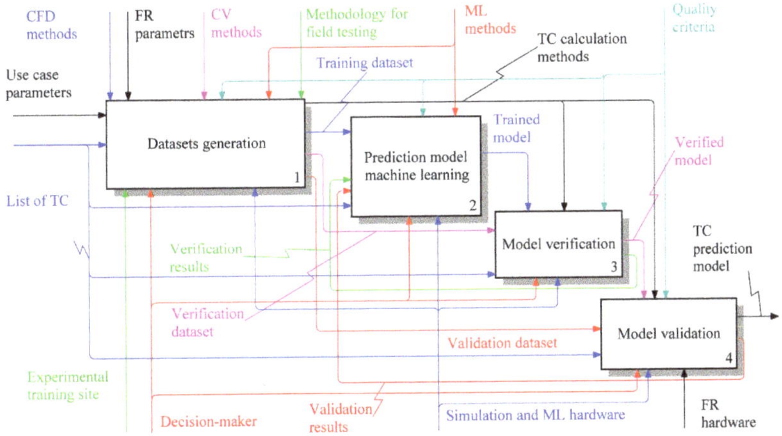

Figure 3. IDEF0 functional diagram for the task of building a model for real-time forecasting of fire-extinguishing agent trajectory.

The generation of initial data (Figure 3, block 1) is based on the usage scenarios of the fire monitor under investigation. Within this work, these scenarios are described by the following set of parameters:

- Ambient air temperature range (T);
- Range of wind load directions (γ—the wind direction angle relative to the horizontal projection of the fire monitor nozzle axis) and velocities (v_w);
- Range of elevation angles for the fire robot monitor's output nozzle (α);
- Range of fire-extinguishing agent flow rates (q_{fea}).

Based on these parameters, sets of scenarios are selected (Figure 4, block 3) for machine learning (Figure 3, block 2), verification (Figure 3, block 3), and validation (Figure 3, block 4) of the trajectory forecasting computational model, adhering to the following principles:

- The sets of training and verification scenarios must not overlap, i.e., scenarios used to check the model's adequacy against results obtained via Computational Fluid Dynamics methods (verification) are not used in its training [26].
- The sets of training and validation scenarios must also not overlap, i.e., scenarios used to check the model's adequacy against results obtained from field tests (validation) are not used in its training.
- The boundaries of the parameter value ranges in the training set must correspond to the usage scenarios of the investigated robotic firefighting systems.
- The validation scenario set can be formed and adjusted based on the results of conducted field tests, as ensuring precise values for the main physical parameters of the experiment is not always feasible.

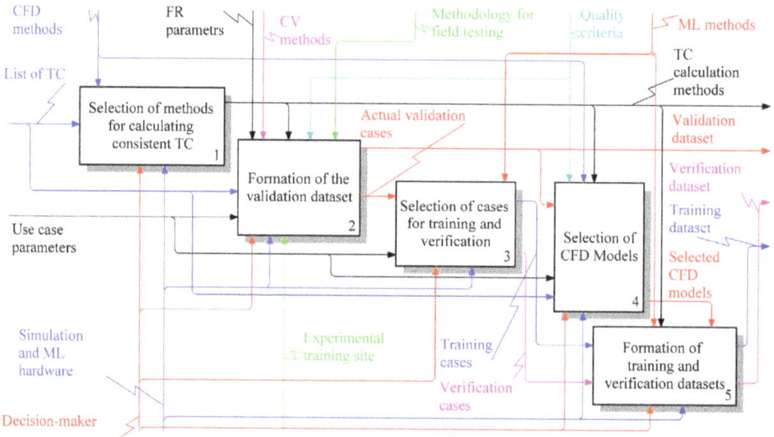

Figure 4. IDEF0 functional diagram for the task of generating initial data for machine learning, verification, and validation of the real-time fire-extinguishing agent trajectory forecasting model.

Based on the formed scenario sets, corresponding datasets are generated (Figure 4, block 5): a training set and a test (verification) set from simulations using CFD methods for modeling the fire-extinguishing agent motion, and an experimental (validation) set (Figure 4, block 2) from field tests. Each such dataset represents a collection of data describing the target parameters of the gas–liquid flows for various values of the main physical parameters. Among the characteristics of greatest interest in this applied field are height, range, upper boundary (trajectory of the leading-edge points) of the fire-extinguishing agent jet, size of the impact area, and displacement of the impact area (Figure 5).

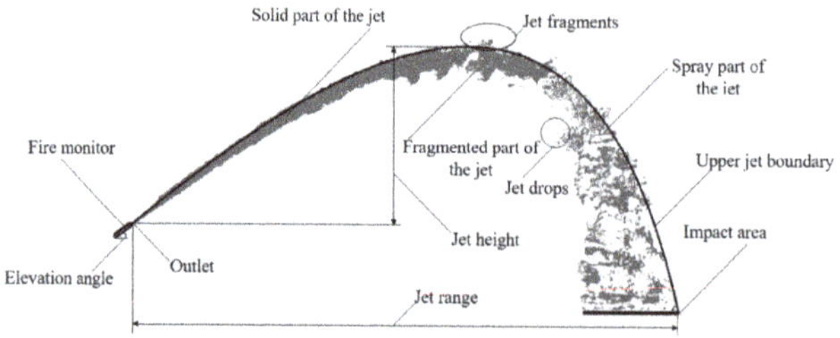

Figure 5. Schematic of a fire-extinguishing agent flow from a robotic fire monitor.

The results of simulating the fire-extinguishing agent flow using CFD methods constitute a large array of data, consisting of numerous physical parameters (coordinates, velocity vectors, density, volume fraction of the substance, etc.) at various points in three-dimensional space and moments in time [27]. To form training and verification datasets from these arrays, and to validate the used models, the alignment of target characteristics is applied (Figure 3, block 1). This alignment ensures that the characteristics describing the dynamics of the investigated flow processes are consistent between the calculation results and the field test results.

Machine learning [28] of the computational model (Figure 3, block 2) is performed based on the constructed training dataset. Its verification (Figure 3, block 3) involves assessing the accuracy of calculating the target characteristics of the fire-extinguishing agent flow against the corresponding parameters obtained from simulations using CFD methods under various scenarios (verification dataset) [29]. Validation (Figure 3, block 4) is performed similarly, but using the results of field tests (validation dataset) [30]. If unsatisfactory quality metrics are obtained (in particular, values of root mean square errors), the model is retrained with different hyperparameters.

2.2. Methodology for Forming Validation Datasets from Field Test Results

Figure 6 presents the functional diagram of the developed technology.

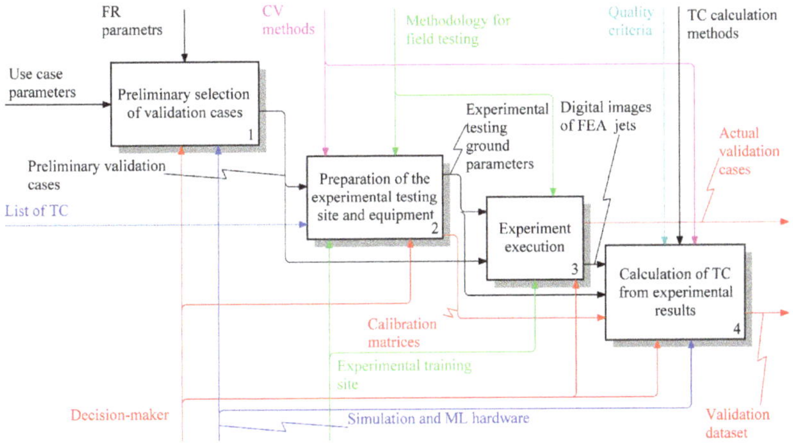

Figure 6. IDEF0 functional diagram of the technology for constructing a 3D model of the fire-extinguishing agent trajectory from field test results.

Preparing the experimental test site and equipment (Figure 6, block 2) also involves camera calibration [31]. This procedure consists of determining their intrinsic parameters, which are subsequently used to perform perspective transformations. Based on the calculated target characteristics (Figure 6, block 4), the accuracy [32] of the constructed three-dimensional models is evaluated by comparing the target characteristics derived from them with those measured during full-scale tests. The specific characteristics compared include the position of the FEA discharge origin, the position of the point on the jet furthest from the origin (based on the outermost droplets), and the position of the jet's highest point.

Preparing the test site (Figure 6, block 2) for investigating fire-extinguishing agent jet trajectories under disturbances involves marking coordinate axes and placing markers at points that serve as reference points for positioning the cameras and the fire monitor under study. The corresponding layout is shown in Figure 7.

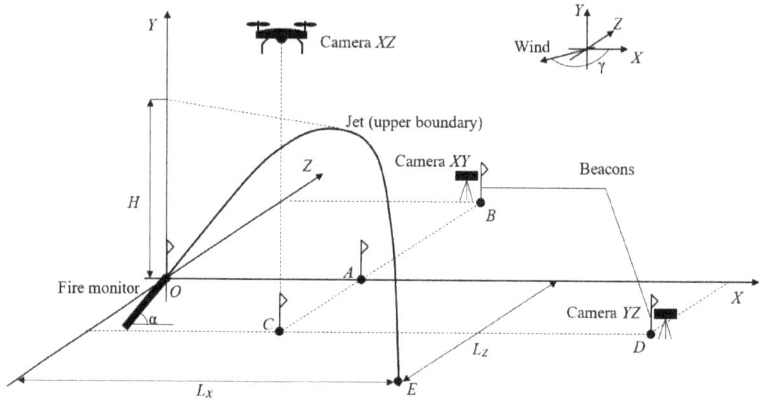

Figure 7. Test site layout.

According to the developed technology, preparing the site for conducting field tests to collect empirical data describing the dynamics of the fire-extinguishing agent flow under disturbances involves solving the following tasks:

1. Preliminary selection of the OX axis location, coordinate origin, and other points, considering the geometric (dimensions, elevation changes) and technical features of the test site, parameters of control (investigated fire monitor elevation angles, fire-extinguishing agent flow rates) and disturbing influences (wind speed and direction, air temperature), characteristics of the digital cameras used, etc.
2. Site marking (determining the positions of points A, B, C, D) using geodetic equipment. At each point (O, A, B, C, D), markers are installed—highly visible vertical objects of known height (e.g., marker flags). These are used for rapid equipment placement and setup (orientation), as well as for refining the geometric parameters of the experiment (characteristic dimensions, camera rotation angles, etc.) based on the acquired digital images.
3. Measuring the coordinates (along OX, OY, OZ axes) of points A, B, C, D based on the site marking results.

Conducting the experiment (Figure 6, block 3) includes three main stages: preparatory work (installation and setup of cameras, the fire monitor under study, etc.), a test jet discharge to verify that the equipment placement corresponds to the parameters of the processes under investigation, and the main phase of field tests. During the main phase, the fire-extinguishing agent jet is recorded in video mode at the maximum possible resolution and frame rate for the equipment used, under various control and disturbance conditions. An audio signal is used for synchronizing the video streams. If necessary, instrumental measurement of various geometric parameters of the jet is performed to assess the accuracy of the results obtained from digital image analysis. The parameters recorded in the test protocol are presented in Table 1.

For constructing the 3D model and calculating target characteristics (Figure 6, block 4) based on digital images formed from the videos obtained during field tests, jet trajectory recognition is performed. This involves constructing arrays of positions for points lying on the upper boundaries of the investigated fire-extinguishing agent flow, within the coordinate systems of each camera's matrix. Boundary extraction of the jets in the images is performed using the algorithm presented in Appendix A. This algorithm is designed to form arrays of coordinates for points on the upper and lower boundaries of the fire-extinguishing agent flow image in the two-dimensional coordinate system of the digital camera's matrix.

The upper boundary is used here and henceforth as the jet trajectory; however, for solving certain tasks, such as estimating the size of the impact area, information about the lower boundary may be required. Therefore, the algorithm for extracting the fire-extinguishing agent flow boundaries provides both curves.

Table 1. Main parameters recorded in the test protocol.

Parameter Name	Measurement Accuracy	Stage of Measurement
coordinates of main site points A, B, C, D, and E (when relevant measurements are taken)	0.1 m—horizontal axes, 0.01 m—vertical axis	site marking (points A, B, C, D), test jet discharge (point E)
deviations along each coordinate axis for the camera XY and YZ locations (matrix center point) relative to points B and D, respectively, rotation angles of cameras XY and YZ relative to horizontal axes	0.1 m—horizontal axes, 0.01 m—vertical axis	preparatory work (after camera installation), test jet discharge (when adjusting camera placement and orientation)
deviations along each coordinate axis of the fire monitor nozzle (FMN) outlet center relative to point O	0.1°	preparatory work (after camera installation), test jet discharge (when adjusting camera placement and orientation)
elevation angle of the FMN output nozzle	0.01 m	main experimental phase (upon each change in FMN elevation angle)
wind direction, speed	0.1°	main experimental phase (upon each change in FMN elevation angle)
air temperature and humidity, atmospheric pressure	determined by the measuring instrument used	main experimental phase (throughout the entire field test period)
working pressure at FMN outlet, temperature, and flow rate of fire-extinguishing agent	determined by the measuring instrument used	main experimental phase (throughout the entire field test period)

To enhance the effectiveness of this algorithm, image processing is performed to delineate the jet boundaries by creating a significant brightness gradient at these boundaries. This operation comprises the following transformations:

- Reduction in brightness in the brightest image regions to prevent loss of detail in subsequent transformations. This is achieved using the methodology presented in [33].
- Local contrast enhancement using the methodology from [34].
- Global contrast enhancement to create a significant brightness gradient between light and dark areas of the image [34].

Under conditions that are complex from the perspective of fire-extinguishing agent boundary extraction, iterative repetition of the aforementioned transformations is possible. Given the appropriate technical capability, it is advisable to conduct full-scale tests during nighttime with the jet illuminated by spotlights. This approach creates a sharp jet boundary due to a significant brightness contrast between the illuminated extinguishing agent and the unlit surrounding space.

Based on the arrays of points constructed from the results of boundary extraction in jet images from different cameras, 3D reconstruction of its trajectory is performed using perspective transformations [35,36]. The corresponding algorithm is presented in Appendix B.

2.3. Methodology for Constructing a Fire-Extinguishing Agent Trajectory from Numerical Simulation Results Consistent with Experimental Data

The results of numerical simulations using CFD methods for each model time step constitute a large dataset describing the state of the investigated space at various points (coordinates X, Y, Z; velocity vector projections Vx, Vy, Vz; volume fractions VF, etc.) [27], typically at the nodes or cells of the computational grid when using grid-based methods for solving the corresponding differential equations [37]. Estimating target characteristics of flow processes in specific regions (jet range and height, impact area, liquid concentration at specified points and their vicinities) from simulation results is generally straightforward. However, in this applied field, it is necessary to compute jet parameters describing their dynamics throughout the entire volume of space they occupy. Constructing these characteristics can be complicated by the geometric dimensions of the simulated flow processes, which significantly hinders the determination of target parameters with a resolution acceptable for the tasks at hand.

Trajectories and streamlines [38] are among the most common characteristics used to describe the dynamics of flow processes along their entire length. However, when investigating fire-extinguishing agent jets, these approaches are not always applicable. For instance, the upper boundary of the jet, determined visually by computer vision systems of robotic installations based on its "densest" region in a given section (corresponding to particles moving with maximum velocities), can differ significantly from curves derived from pathlines and streamlines.

Using threshold values for the liquid volume fraction to exclude spray fragments from characteristic calculations is ineffective because a significant portion of the jet downstream, due to breakup, has small VF values, yet these must be accounted for when constructing the flow boundary. It is also worth noting that when using the Volume of Fluid (VOF) method [39] for motion modeling with relatively large cells to reduce computation time, the resulting streamlines in the spray part of the jet can extend substantially beyond its actual boundaries.

Considering these problems for tasks related to assessing flow characteristics based on visualized data from field tests and observations (machine learning, verification, and validation of used models), a method for constructing the line of maximum flow velocities was developed. This line is defined by fragments of the substance moving with the maximum velocities and corresponds to the visually "densest" part of the jet, upon which its upper boundary is formed. The proposed method for constructing the fire-extinguishing agent jet trajectory based on numerical simulation results is implemented by the algorithm presented in Appendix C.

Figure 8 presents the lines of maximum velocities and streamlines constructed from the simulation results of fire-extinguishing agent motion from a fire monitor under crosswind, using the k-ε turbulence model and the Volume of Fluid (VOF) method. The dots in Figure 8 denote computational grid nodes where the volume fraction of the fire-extinguishing agent exceeds a specified threshold value. Darker dots correspond to moving jet fragments, while lighter ones correspond to stationary fragments settled on the ground. The lines of maximum velocities are depicted by red curves, and the streamlines by black curves.

The line of maximum velocities envelopes the main part of the simulated flow. The volume fraction values in grid nodes located outside this curve are negligible, meaning this line forms the boundary of the free jet. In contrast, the streamline passes closer to its central part and, in the spray region, extends beyond its limits (Figure 8). This behavior is explained by the fact that when constructing the streamline, cells where the liquid volume fraction is non-zero are considered. However, in the far-field region of the jet, sprayed fragments of the fire-extinguishing substance move almost vertically downward under gravity, while adjacent air regions, accelerated by the main flow and containing insignificant amounts of liquid, continue their predominantly horizontal motion. Therefore, the corresponding curve, constructed considering them, extends beyond the free jet boundary. It is important to note that the proposed method does not have this drawback.

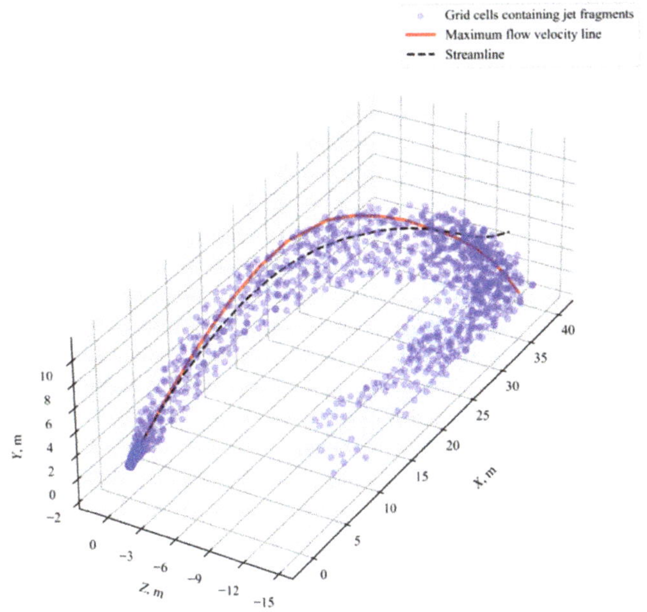

Figure 8. Results of constructing the line of maximum velocities and the streamline.

2.4. Selection of Computational Fluid Dynamics Methods for Simulating Fire-Extinguishing Agent Motion Under Wind Influence

The investigated jets are characterized by large geometric trajectory dimensions (tens of meters), high initial velocities (tens of meters per second), breakup and atomization of the continuous flow into separate fragments and droplets of various sizes (Figure 5), as well as reverse processes like droplet coalescence (e.g., in the impact area), which significantly increases the computational complexity of the simulation task. Specifically, the mathematical description of these phenomena requires the use of appropriate methods and computational grid cell sizes for discretizing the investigated space. Considering the height and range of the studied liquid flows, this demands substantial computational resources.

From a numerical modeling perspective, the investigated free jets represent multiphase flows [40]: air (primary phase)—fire-extinguishing agent (secondary phase). If the characteristic size of the interface significantly exceeds the corresponding computational grid cell sizes in the region of the continuous jet (Figure 5), the Volume of Fluid (VOF) method [39] can be used for its description, which reduces computation time compared to the Eulerian model [41]. This can be relevant for tasks requiring trajectory construction across a wide range of parameter values influencing the investigated dynamics, particularly for generating datasets for training and verifying neural network models.

However, in cases where liquid breakup and atomization (Figure 5) significantly influence its trajectory, it is necessary to account for not only its continuous phase but also its dispersed phase [42], and the corresponding transitions between them. Various approaches are used for this purpose, specifically, hybrid methods jointly using two models, for example, VOF-DPM [43]: VOF for the continuous phase and DPM (Lagrangian model) for the discrete phase. This technology allows accounting, within the scope of the considered task, only for jet breakup and atomization. In studies where the reverse transition (coalescence of individual droplets, etc.) is a subject of investigation, more complex models based on the Eulerian approach [44–46] are applied. These approaches possess significant computational complexity, which can substantially increase calculation time, especially within the geometric scales characteristic of this research's application field. Furthermore, assessing the degree of influence of phase transitions on the trajectory of the free jet, which is the control object (particularly in tasks of directing the fire-extinguishing agent flow from a robotic firefighting complex monitor to a specified spatial region), presents a rather complex problem. Therefore, the selection of computational models (Figure 4, block 4) and their parameters for solving the stated task is performed to ensure high accuracy and calculation speed relative to the tasks being solved (Figure 9).

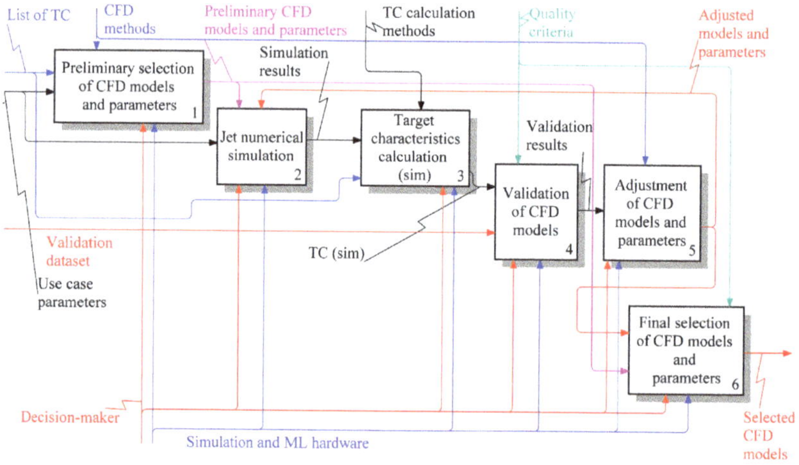

Figure 9. Functional diagram of the technology for selecting Computational Fluid Dynamics methods and their parameters for simulating fire-extinguishing agent motion.

The selection of models and their parameters for describing free fire-extinguishing agent jets, aiming to meet one of the quality criteria (acceptable calculation speed), proceeds from computationally simplest to more complex, i.e., those requiring the solution of a greater number of equations. An important feature of the target flow characteristics identified earlier is the absence of the need to account for transitions from the dispersed phase back to the continuous phase. This allows the use of the simplest preliminary CFD models (Figure 9, block 1): k-ε [47], k-ω [48] for describing turbulence, and VOF [39] for describing phases. Calculation time is also influenced by factors such as the geometric dimensions of the investigated space, the number of computational grid cells, boundary conditions, and the use of additional functions (e.g., for damping or limiting turbulence).

Validation (Figure 9, block 4) of the CFD models used in this study was performed by assessing the deviations (accuracy) of the calculation results (Figure 9 blocks 2, 3) relative to those obtained from field tests with corresponding initial parameter values. For this purpose, the used target characteristics were aligned (Table 2).

The simulation accuracy, performed in Ansys Fluent [49], compared to experimentally obtained results, is assessed within the presented framework based on the relative deviation (1) for simple characteristics (jet height, range, impact area displacement) and the root mean square deviation (2) for complex ones (jet upper boundaries).

$$\mathrm{RD} = |X_e - X_m| / X_e \cdot 100\%, \tag{1}$$

where $X_e \neq 0$, X_m are the values of the simple target characteristic obtained from experimental and simulation results, respectively.

$$\text{RMSD} = \sqrt{\frac{1}{N} \cdot \sum_{i=1}^{N} \left| P_{e,i} - P_{m,i} \right|^2} \tag{2}$$

where N is the number of points of the jet upper boundary constructed from experimental results,

$P_{e,i}$ are the coordinates of the i-th point of the jet upper boundary constructed from experimental results,

$P_{m,i}$ are the coordinates of the point on the jet upper boundary constructed from simulation results closest to point $P_{e,i}$.

Table 2. Alignment of target characteristics determined from simulation and experimental results.

Target Characteristic Name	Evaluation Method from Simulation Results	Evaluation Method from Experimental Results
jet height	vertical distance from the fire monitor outlet to the highest point of the line of maximum velocities	distance Y in Figure 7, computed based on the jet upper boundary constructed from experimental results
jet range	horizontal distance (parallel to the projection of the output nozzle axis onto the horizontal plane) from the fire monitor outlet to the farthest point on the line of maximum velocities	distance L_X measured during the experiment (Figure 7)
impact area displacement	horizontal distance (perpendicular to the projection of the output nozzle axis onto the horizontal plane) from the fire monitor outlet to the farthest point on the line of maximum velocities	distance L_Z measured during the experiment (Figure 7)
jet upper boundary	line (set of points) of maximum fire-extinguishing agent flow velocities (Appendix C)	line (a set of points) generated by the 3D jet trajectory reconstruction algorithm (Appendix B)

The decision to change (complexify) the CFD models or their parameters is made depending on the values of the specified deviations. In this study, the fulfillment of conditions (3)–(4) was used as criteria for achieving acceptable accuracy:

$$\text{RD} \leq 10\%, \tag{3}$$

$$\text{RMSD} \leq 1 \text{ м} \tag{4}$$

If these conditions are not met, correction (Figure 9, block 5) of the used models or their parameters is performed, followed by repeated execution of numerical simulation, target characteristic calculation, and validation operations.

The primary selection of CFD models and their parameters (Figure 9, block 1) was performed for the task of simulating the water jet trajectory from an LS-P20U fire monitor with a high flow rate of fire-extinguishing agent without accounting for wind, to accelerate calculations through model simplification. Validation was based on the jet's range, height, and upper boundary trajectory. According to the study results, it was established that the computational model based on k-ε (in the RNG modification with turbulence damping) and VOF (without accounting for implicit forces, with a liquid volume fraction limit of 10^{-6}) satisfies the specified accuracy criteria and also provides the shortest computation time among all investigated configurations. The results of the preliminary model selection are presented in Table 3 (the computation time is specified for a personal computer with an AMD Ryzen 3 2200U processor).

As follows from the obtained results, the trajectories satisfying conditions (3)–(4) are those constructed based on the combined use of the k-ε and VOF models with various tuning parameters. The minimum computation time is provided by the scheme with a liquid volume fraction limit of 10^{-6} and without the inclusion of implicit forces; therefore, this scheme was selected as the baseline for further calculations. Figure 10 shows fragments of images of the water jet from the LS-P20U, obtained for the following control parameter values: output nozzle elevation angle 57°, fire-extinguishing agent flow rate 10.2 l/s.

Table 3. Preliminary model selection results.

Models and Parameters	Computation Time, h	RD (Range), %	RD (Height), %	RMSD, m
k-ε RNG, turbulence damping, VOF with liquid volume fraction limit of 10^{-8}	11	3.9	7.8	0.29
k-ε RNG, turbulence damping, VOF with liquid volume fraction limit of 10^{-6}	10	4.4	5.2	0.27
k-ε RNG, turbulence damping, VOF with liquid volume fraction limit of 10^{-6}, implicit body force	11	4.4	5.2	0.27
k-ω SST, turbulence damping, VOF with liquid volume fraction limit of 10^{-6}	11	6.5	10.2	0.48
k-ε RNG, Eulerian model	12.5	10.1	12.4	0.76

Figure 10. Fragments of the fire-extinguishing agent jet image from the fire monitor after preliminary processing (**left**) and boundary extraction (**right**).

The left part of Figure 10 shows a fragment of a digital photograph of the fire-extinguishing agent flow after preliminary processing, and the right part shows the result after extracting the upper and lower boundaries based on the algorithm described in Appendix A. Figure 11 shows superimposed curves corresponding to the jet upper boundaries constructed from experimental and simulation results, and Table 4 presents the validation results.

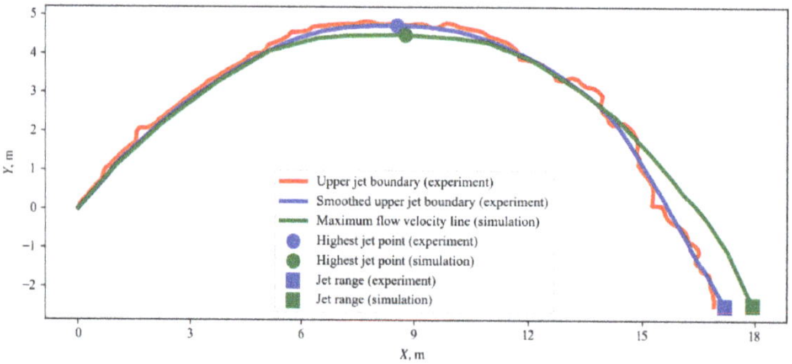

Figure 11. Upper jet boundaries obtained from experimental and simulation results.

Table 4. Assessment of simulation accuracy relative to experimental results.

Target Characteristic Name	Experimental Value, m	Simulation Value, m	Deviation RD/RMSD	Fulfillment of Conditions (3)–(4)
jet range	17.2	17.9	4.4%	$4.4\% \leq 10\%$
jet height	4.7	4.5	5.2%	$5.2\% \leq 10\%$
jet upper boundary	–	–	0.27 m	$0.27\text{ m} \leq 1\text{ m}$

As follows from Table 4, conditions (3)–(4) are met for all presented characteristics, which corresponds to acceptable model accuracy. The calculation time was 10 h (using a personal computer with an AMD Ryzen 3 2200U processor), which also satisfies the specified quality criteria. The validation results of the selected models for calculating the water jet trajectory from the fire monitor accounting for crosswind are presented in Section 3.2.

2.5. Experimental Evaluation of the Real-Time Water Jet Trajectory Prediction Methodology Effectiveness

The effectiveness of the developed prediction methodology was evaluated using a water jet from the LS-P20U fire monitor, which serves as the actuator for a firefighting robot. For this purpose, field tests were conducted under a limited set of control and disturbance scenarios, described by the following parameter values:

- Elevation angles α, °: [30, 44.5, 57]—based on actual measurements during experiments (planned values: [30, 45, 60]). The selection of these values is determined by the limitations of the experimental test site and the coverage of the maximum (α = 30°) and medium range (α = 60°) of the jet.
- Fire-extinguishing agent flow rates q_{fea}, l/s: [12.1, 16.1, 17.1, 39.8]—based on actual measurements during experiments (planned values: [12, 16, 18, 40]). The selection is determined by the constraints of the experimental test site, the equipment used, and the coverage of the maximum (q_{fea} = 40 l/s) and short-range (q_{fea} = 12 l/s) jet characteristics.
- Wind direction angles γ, °: [65, 76, 96, 104].
- Wind speed v_w, m/s: [0, 1.4, 2, 2.5, 3.3].
- Air temperature T, °C: [−2.2, −1.4, 2.5].

The experimental values of the disturbance parameters are limited by the actual weather conditions at the test site during the full-scale tests. Generating validation samples for a broader range of external conditions, such as high wind speeds, requires conducting corresponding experiments, which could be a subject of further research.

Based on the obtained results, a dataset was generated and used for validating the selected CFD models and the trained neural network model for water jet trajectory prediction (a multilayer perceptron—MLP with error backpropagation using the scikit-learn library [50]). The MLP was chosen as a preliminary baseline model due to its simpler architecture compared to other neural networks, allowing an assessment of the approach's effectiveness for the defined task within the limited set of scenarios. The primary advantages of the MLP in this context include simplicity of implementation; training and inference speed; ease of identifying potential issues; and its utility as a starting model for a progressive complexity strategy, potentially evolving into more advanced architectures such as physics-informed neural networks (PINNs) or neural ensembles tailored for different scenarios. The main disadvantage of an MLP for this task is the potential for a critical reduction in prediction accuracy for scenarios significantly different from those in the training set. Therefore, during the training data preparation stage, it is necessary to ensure coverage of potential operational scenarios, which may require a large dataset.

Table 5 presents a comparison of the MLP with other architectures [51] in terms of the most significant potential limitations and performance differences relevant to the task.

Table 5. Comparison of potential limitations and performance of architectures.

Architecture	Potential Limitations	Performance
MLP	large training dataset requirement, difficulty with extrapolation beyond training data	high
RNN	high complexity for long trajectory	potentially reduced for very long trajectories
PINN	high complexity in defining governing physical equations	high
NNGP	high computational complexity, difficult implementation	potentially substantially higher than MLP

To overcome the mentioned drawbacks of the MLP, promising directions within the progressive complexity strategy include transition to PINNs via hybridization [51] (adding physics-based constraints through regularization), ensemble methods [52] (combining multiple MLPs for different scenario groups), and others.

The machine learning and verification of the prediction model were performed using datasets generated from simulation results based on the selected CFD methods. The scenarios used for this purpose are described by the following parameter values:

- Elevation angles α, °: [30, 35, ..., 60].
- Fire-extinguishing agent flow rates q_{fea}, l/s: [12, 14, ..., 20, 25, ..., 40].
- Wind direction angles γ, °: [60, 65, ..., 120].
- Wind speed v_w, m/s: [0, 0.5, ..., 4].
- Air temperature (fixed) $T = 0$ °C.

The values within these ranges were selected based on the following principles: coverage of the parameters from the scenarios used to build the validation dataset; and a parameter value increment ensuring a sufficient size for the training and verification datasets while maintaining an acceptable total simulation time. The total number of scenarios generated in this study, based on combinations of the parameter ranges for control (α, q_{fea}) and disturbance (γ, v_w) inputs listed above, was 735, with 100 trajectory points per scenario. To reduce the total model preparation and computation time, simulations (run on a computational cluster) for each elevation angle value were executed with varying values of the other parameters. These parameters were changed sequentially only after a stable extinguishing agent trajectory was achieved.

The training (P_{LEARN}) and verification (P_{VER}) datasets were generated from simulation results (P_m) in an 80/20 volume ratio. The corresponding dataset sizes were 588 and 147 scenarios, respectively (with 100 trajectory points per scenario). The loss function used for machine learning of the prediction model was the root mean square deviation (RMSD):

$$\text{RMSD} = \sqrt{\frac{1}{N} \cdot \sum_{i=1}^{N} \left| P_i - P_{\text{MLP},i} \right|^2}, \quad (5)$$

where

- N is the number of points in the jet trajectory constructed by the trained model;
- $P_{\text{MLP},i}$ are the coordinates of the i-th point of the jet upper boundary constructed by the trained model;
- P_i are the coordinates of the point in the jet trajectory (constructed from simulation or experimental results) closest to $P_{\text{MLP},i}$ ($P = P_{\text{LEARN}}$ for the training dataset, $P = P_{\text{VER}}$ for the verification dataset, $P = P_e$ for the validation dataset).

Hyperparameter tuning [53,54] was conducted to ensure the required quality of target parameter calculation, convergence, and protection against overfitting. The criterion for stopping the increase in neural network model complexity was the fulfillment of condition (4) for the RMSD relative to the verification dataset. The final hyperparameters of the trained model are presented in Table 6.

Table 6. Final hyperparameters of the trained model.

Hyperparameter Name	Value
number of hidden layers	3
neuron activation function	tanh
L2 regularization parameter	0.001
solver for weight optimization	Adam
validation fraction	0.15
initial learning rate	0.001
learning rate	adaptive
early stopping	enabled
size of minibatches	250

The results of the trained model's accuracy assessment are presented in Section 3.3.

3. Results

3.1. Reconstruction of Water Jet Trajectory from a Fire Monitor Based on Field Test Results

To evaluate the effectiveness of the developed technology for 3D reconstruction of fire-extinguishing agent jets, field tests were conducted at the experimental site. Table 7 presents the coordinates and rotation angles of the cameras, corresponding to a jet discharge with the following characteristics: LS-P20U output nozzle elevation angle $\alpha = 44.5°$; fire-extinguishing agent flow rate q_{fea} = 16.1 l/s; wind speed v_w = 2 m/s; wind direction angle $\gamma = 104°$.

Table 7. Camera parameters.

Parameter Name	Camera XY	Camera YZ	Camera XZ
Coordinates (X; Y; Z), m	(23.4; −1.38; 30)	(52.1; −1.8; −0.3)	(14; 34.9; 0)
Rotation Angles (OX; OY; OZ), °	(20; 0; 0)	(0; −1.2; 20)	(0.5; 0; 0)

Figure 12 shows fragments of the original images, obtained at the same moment in time from sound-synchronized video recordings from three cameras, along with the corresponding results of their preliminary processing and jet boundary extraction.

Figure 13 presents the results of constructing the 3D model of the jet trajectory (original and smoothed curves) using the algorithm described in Appendix B, as well as control points derived from it and from measurements: the jet origin (center of the monitor outlet), the jet's highest point, and the point on the jet farthest from the output nozzle.

Table 8 presents a comparative analysis of the main geometric characteristics of the jet, obtained from the constructed model and from measurements taken during the field tests.

Table 8. Comparative analysis of jet geometric characteristics.

Target Characteristic Name	Measured Value, m	Model Value, m	Absolute Error, m	Relative Error, %
jet range	25.2	24.9	0.3	1.2
jet height	7.65	7.37	0.28	3.7
impact area displacement	3.58	3.57	0.01	0.3

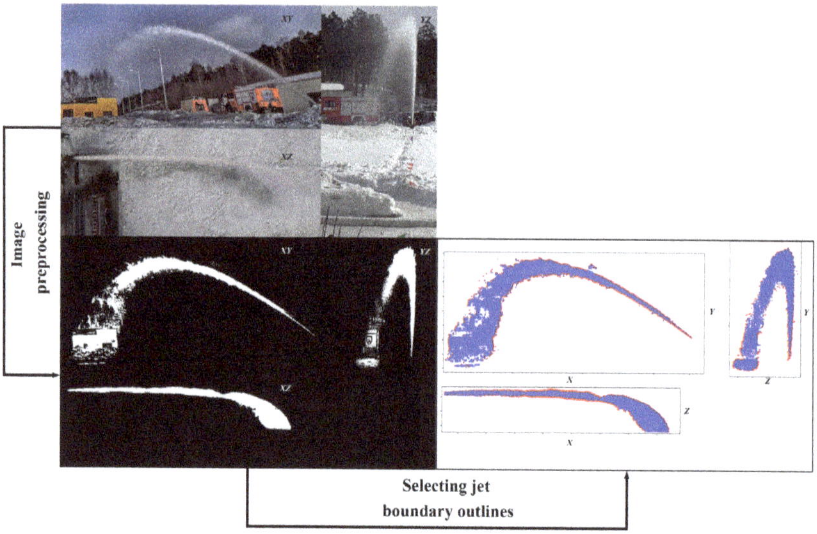

Figure 12. Fragments of original images, results of their preliminary processing, and jet boundary extraction.

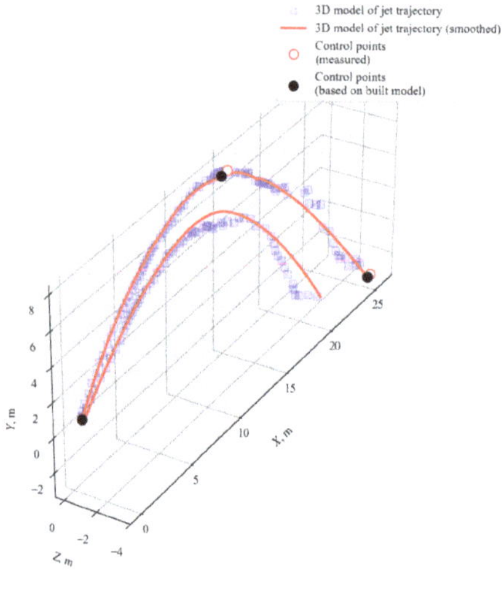

Figure 13. Results of constructing the 3D model of the jet trajectory.

The deviations of the model positions for the highest point and the point farthest from the fire monitor (Figure 3) from the measured ones do not exceed 1 m, and the corresponding origin coordinates coincide. The maximum relative error among the characteristics presented in Table 8 was 3.7%, indicating the sufficiently high accuracy of the proposed technology.

To assess the sensitivity of the developed methodology to errors in determining camera position and orientation angles, jet images were generated from the smoothed trajectory (upper boundary) using forward perspective transformations [35]. Subsequently, corresponding models were constructed using the proposed technology with various deviations of the cameras' geometric parameters (coordinates in the test site system and rotation angles about coordinate axes) from the values used to generate the images. Accuracy was assessed based on the root mean square deviation [32] relative to the original trajectory. The results are presented in Figure 14.

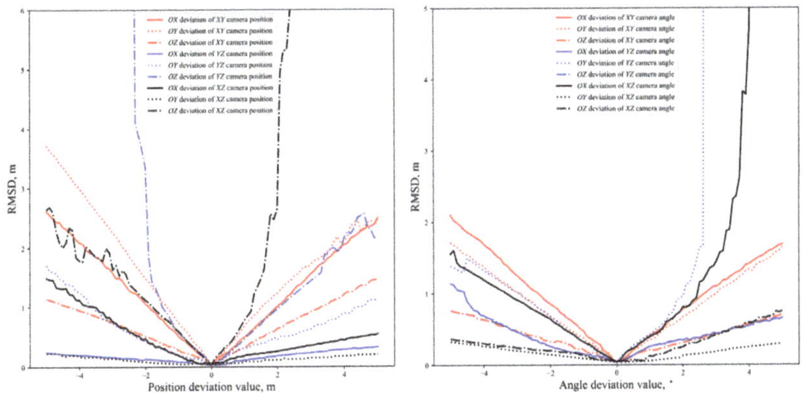

Figure 14. Dependence of model construction accuracy on errors in determining camera position and rotation angles.

As can be seen from Figure 14, the least significant impact on trajectory reconstruction accuracy is exerted by errors in determining the positions and rotation angles of cameras relative to axes perpendicular to the planes of their sensor matrices (in the initial position): deviation along the OZ axis for the XY camera; deviation along the OX axis for the YZ camera; deviation along the OY axis for the XZ camera. The most significant impact on accuracy is attributed to the following errors: OY deviation of position, OX deviation of angle for the XY camera; OZ deviation of angle for the YZ cam-era; OZ deviation of position, OX deviation of angle for the XZ camera.

It is worth noting that errors in determining camera coordinates and orientation within 10–20 cm and 0.5°, respectively, can be considered insignificant in the considered example.

3.2. Validation Results of Computational Fluid Dynamics Models

Figures 15 and 16 and Table 9 present the results of the accuracy assessment for the CFD methods and their parameters selected in Section 2.4 (k-ε in the RNG modification with turbulence damping and VOF without accounting for implicit forces, with a liquid volume fraction limit of 10^{-6}) when simulating the fire-extinguishing agent flow from the LS-P20U monitor for the scenario: α = 44.5°, q_{fea} = 12.1 l/s, v_w = 1.4 m/s, γ = 96°.

Table 9. Assessment of simulation accuracy relative to experimental results.

Target Characteristic Name	Experimental Value, m	Simulation Value, m	Deviation RD/RMSD	Fulfillment of Conditions (3)–(4)
jet range	17.7	18.3	3.4%	3.4% ≤ 10%
jet height	6.5	6.4	1.5%	1.5% ≤ 10%
impact area displacement	4.2	4.0	4.7%	4.7% ≤ 10%
jet upper boundary	–	–	0.29 m	0.29 m ≤ 1 m

As follows from Table 9, conditions (3)–(4) are met for all presented characteristics, which corresponds to acceptable model accuracy. The calculation time for 10 s of fire-extinguishing agent flow motion under wind influence was 10 h, which also satisfies the specified quality criteria.

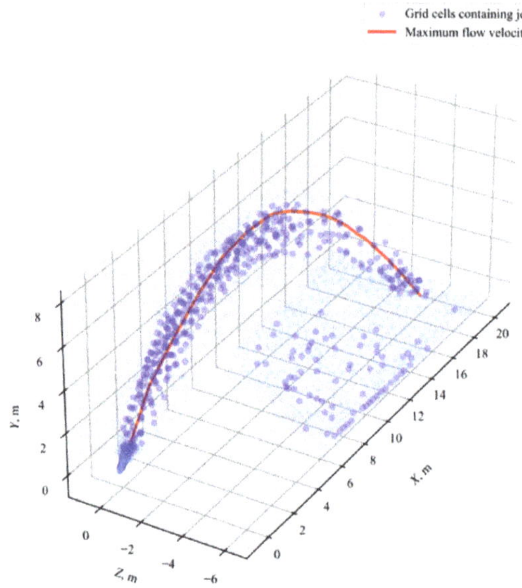

Figure 15. Simulation results and flow characteristics constructed based on them.

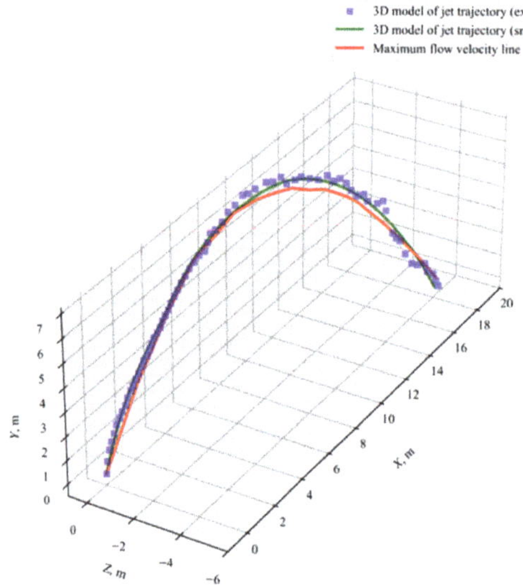

Figure 16. Jet trajectories obtained from experimental results and from simulation accounting for wind.

3.3. Results of Experimental Evaluation of the Real-Time Water Jet Trajectory Prediction Methodology Effectiveness

Figure 17 and Table 10 present the results of constructing the water jet trajectory from the LS-P20U fire monitor for the scenario: $\alpha = 44.5°$, $q_{fea} = 16.1$ l/s, $v_w = 2$ m/s, $\gamma = 104°$, constructed from field test results and based on the trained model.

The deviation of the predicted trajectory from the experimental one is insignificant; specifically, the root mean square deviation along its entire length is 0.48 m, indicating sufficiently high accuracy. Furthermore, the prediction time on a personal computer with an AMD Ryzen 3 2200U processor was 0.009 s, which is several orders of magnitude less than the period required for jet stabilization and thus corresponds to real-time operation.

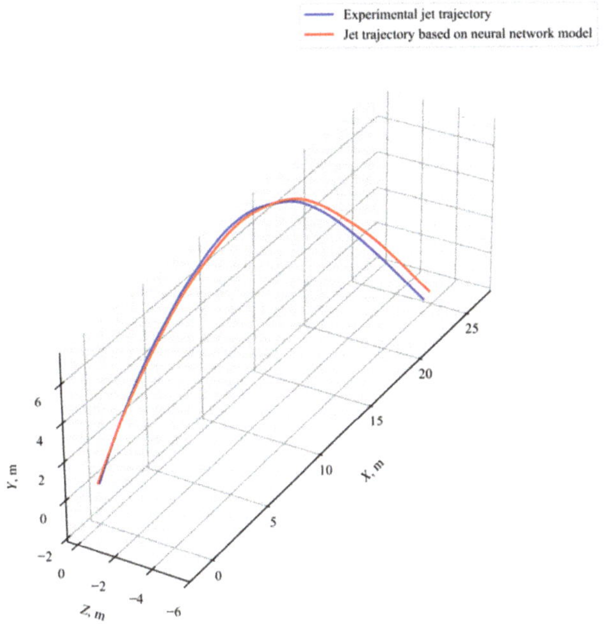

Figure 17. Jet trajectories obtained from experimental results and based on the trained model.

Table 10. Assessment of trajectory prediction accuracy relative to experimental results.

Target Characteristic Name	Validation Value, m	Predicted Value, m	Deviation RD/RMSD	Fulfillment of Conditions (3)–(4)
jet range	24.9	25.8	3.6%	$3.6\% \leq 10\%$
jet height	7.37	7.21	2.2%	$2.2\% \leq 10\%$
impact area displacement	3.57	3.48	2.5%	$2.5\% \leq 10\%$
jet upper boundary	–	–	0.48 m	$0.48 \text{ m} \leq 1 \text{ m}$

Table 11 presents information about the errors of the constructed neural network model relative to the training, verification (test), and validation datasets.

Table 11. Assessment of training accuracy.

Dataset Name	Minimum RMSD Value, m	Maximum RMSD Value, m	Mean RMSD Value, m
training	0.06	0.48	0.12
verification	0.06	0.51	0.14
validation	0.22	0.71	0.52

As follows from Table 11, the maximum prediction error based on the trained model does not exceed 0.71 m, which meets the accuracy criterion (4) formulated above.

4. Discussion

The presented methodology for the combined use of machine learning and Computational Fluid Dynamics methods for predicting the trajectory of fire-extinguishing agent from a fire monitor allows the stated task to be solved with high accuracy and speed corresponding to real-time operation. This, in turn, enables its application for robotic control of jet targeting towards a specified goal under complex conditions (wind, poor visibility due to smoke).

It should be noted that generating the training and test (verification) datasets based on CFD methods solves the problem associated with the impossibility of conducting field tests across a wide range of disturbances. Thus, after training and verifying the predictive model, its validation based on the experimental dataset is conducted on a limited set of fire robot usage scenarios, which can be formed over an extended period, for instance, as relevant weather conditions (air temperature, wind direction and speed) occur.

The results presented in the article represent a preliminary assessment of the proposed methodology's effectiveness, and their applicability is limited to the investigated fire monitor nozzle and the specified set of scenarios. Specifically, the development and testing of the neural network model were conducted under fixed values of air temperature, extinguishing agent temperature, jet spray angle, etc. Strong wind conditions (outside the studied scenarios) were also not considered. Increasing the number of parameters and significantly expanding their value ranges will necessitate additional experiments and increased model complexity, which, in turn, may lead to a reduction in prediction accuracy and computational speed. To solve this problem, more efficient architectures can be used: physics-informed neural networks [51,55], including those based on Kolmogorov–Arnold networks [56,57], neural network ensembles [58], etc.

A further direction for the development of this research is to evaluate the effectiveness of the developed technology on more complex neural network architectures and an expanded set of scenarios. This expanded set should cover a wide range of elevation angles, including negative values, jet spray angles, wind direction, and speed angles, as well as ambient air temperatures. It is worth noting that including scenarios corresponding to extremely low temperatures (arctic conditions) will require accounting for the influence of phase transitions, leading to increased complexity of the used CFD models. Furthermore, the application of the methodology to other fire-extinguishing agents is of high research interest, specifically for foam of various expansion ratios, whose motion dynamics differ significantly from those of water.

5. Patents

The method for predicting the trajectory of a fire-extinguishing agent jet from a fire monitor under wind influence based on a neural network model and mathematical modeling depicted in Section 2.1 had patents awarded with registered number RU2830398C1. The method for constructing a three-dimensional model of the trajectory of a fire-extinguishing agent jet from a fire monitor depicted in Section 2.2 had patents awarded with registered number RU2843123C1.

Author Contributions: I.P.—conceptualization, methodology, experiment, simulation, machine learning, validation, writing. S.C.—administration, writing, revision and proofreading. All authors have read and agreed to the published version of the manuscript.

Funding: This research received no external funding.

Data Availability Statement: The data are contained within the article.

Acknowledgments: The authors would like to thank the editor and reviewers for their helpful comments and suggestions. DeepSeek-V3.2 was used for the purpose of checking and improving the quality of the English language. The authors have reviewed and edited the output and take full responsibility for the content of this publication.

Conflicts of Interest: The authors declare no conflicts of interest.

Abbreviations

The following abbreviations are used in this manuscript:

CFD	Computational fluid dynamics
ML	Machine learning
FMN	Fire monitor nozzle
RNG	Re-normalization group
VOF	Volume of fluid
RMSD	Root mean square deviation
RD	Relative deviation
TC	Target characteristics
PINN	Physics-informed neural network
RNN	Recurrent neural network
NNGP	Neural network gaussian process
MLP	Multi-layer perceptron
UAV	Unmanned aerial vehicle
FEA	fire-extinguishing agent

Appendix A

Algorithm for Fire-Extinguishing Agent Flow Boundary Extraction

Input Data:

- Arrays of coordinates for multiple points of the fire-extinguishing agent flow image: x, y.
- Angular resolution of the algorithm: $\Delta\varphi$.
- Radial resolution of the algorithm: Δr.
- Maximum allowable increase in the polar radius between adjacent points of the fire-extinguishing agent flow boundary image: ΔR.
- Output Data:
- Arrays of coordinates for points of the upper and lower boundaries of the fire-extinguishing agent flow image: xu, yu, xd, yd.

Step 1. Reset the arrays of coordinate points for the upper and lower boundaries of the fire-extinguishing agent flow image: xu, yu, xd, yd.

Step 2. Determine the position of the polar coordinate system center (Figure A1):

$$\begin{aligned} px &= \frac{x_{\min}+x_{\max}}{2} \\ py &= y_{\min} \end{aligned} \tag{A1}$$

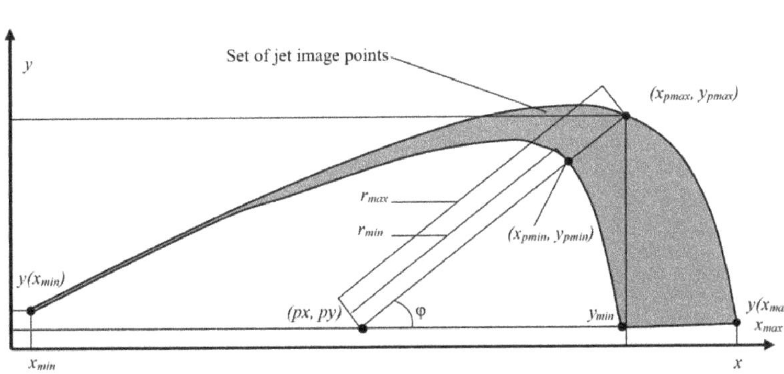

Figure A1. Explanatory diagram for the algorithm of primary fire-extinguishing agent flow boundary contour extraction.

Step 3. Calculate the limits of variation for the polar angle and radius encompassing the fire-extinguishing agent flow image:

$$\begin{aligned}\varphi_1 &= \arccos\left(\frac{x_{\min}-px}{\sqrt{(x_{\min}-px)^2+(y(x_{\min})-py)^2}}\right)\\ \varphi_2 &= \arccos\left(\frac{x_{\max}-px}{\sqrt{(x_{\max}-px)^2+(y(x_{\max})-py)^2}}\right)\end{aligned} \quad \text{(A2)}$$

$$\begin{aligned}r_1 &= \min\left(\sqrt{(x-px)^2+(y-py)^2}\right)\\ r_2 &= \max\left(\sqrt{(x-px)^2+(y-py)^2}\right)\end{aligned} \quad \text{(A3)}$$

Step 4. Begin outer loop (over polar angles): $\varphi = \varphi_1$.

Step 5. Calculate the parameters of the line equation for the section:

$$\begin{aligned}&A \cdot x + B \cdot y + C = 0\\ &A = -\sin(\varphi)\\ &B = \cos(\varphi)\\ &C = -(A \cdot px + B \cdot py)\end{aligned} \quad \text{(A4)}$$

Step 6. Calculate the distances from the fire-extinguishing agent flow image points to the section line:

$$d(x,y) = |A \cdot x + B \cdot y + C| \quad \text{(A5)}$$

Step 7. Populate arrays xs, ys with points lying in the vicinity of the section line: $(xs, ys) = (x, y) : dx(x, y) \leq \Delta r$.

Step 8. If arrays xs, ys are not empty, then determine the radius vectors and coordinates of the upper and lower boundary points of the fire-extinguishing agent flow image in the current section (Figure A1):

$$\begin{aligned}r_{\max} &= \max(\sqrt{(xs-xp)^2+(ys-yp)^2})\\ r_{\min} &= \min(\sqrt{(xs-xp)^2+(ys-yp)^2})\\ x_{p\max} &= px + r_{\max} \cdot \cos(\varphi)\\ y_{p\max} &= py + r_{\max} \cdot \sin(\varphi)\\ x_{p\min} &= px + r_{\min} \cdot \cos(\varphi)\\ y_{p\min} &= py + r_{\min} \cdot \sin(\varphi)\end{aligned} \quad \text{(A6)}$$

The coordinates of the boundary points are added to the arrays xu, yu, xd, and yd.

Step 9. Calculate the limits of variation for the polar radius:

$$\begin{aligned}r_1 &= r_{\min} - \Delta R\\ r_2 &= r_{\max} + \Delta R\end{aligned} \quad \text{(A7)}$$

Step 10. If $\varphi > \varphi_2$, then $\varphi = \varphi - \Delta\varphi$ and go to Step 5.

Step 11. The algorithm returns arrays with coordinates of the primary upper and lower boundary points of the fire-extinguishing agent flow image: xu, yu, xd, and yd.

Appendix B

Algorithm for Constructing a 3D Model of the Jet Trajectory under Disturbances

An explanatory diagram for the algorithm, illustrating the transition between its points, is presented in Figure A2.

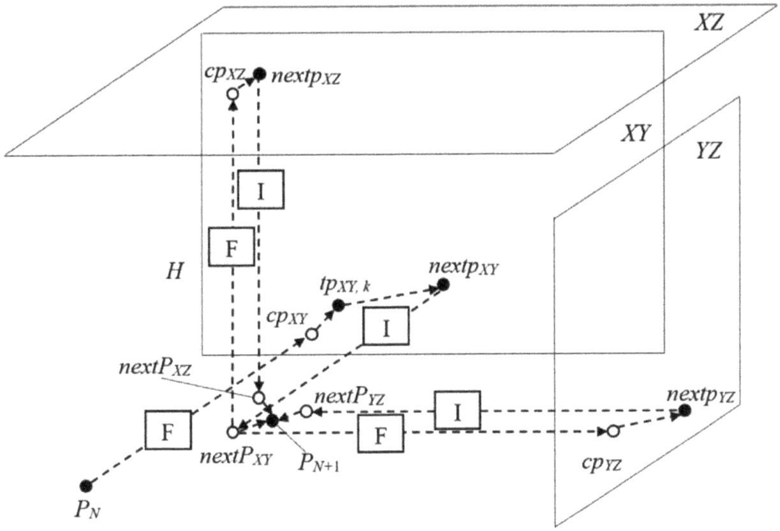

Figure A2. Explanatory diagram for the algorithm for constructing a 3D model of the fire-extinguishing agent flow trajectory (F—forward perspective transformation operation; I—inverse perspective transformation operation).

Input Data:

- Array of digital camera indices: $Cam = \langle m \rangle$, where m corresponds to the coordinate planes to which the camera matrices used for jet imaging are parallel (XY, YZ, ZX).
- Arrays of coordinates of jet trajectory image points from various cameras: tp_m. Each point in the array is described by homogeneous coordinates [25]: $\begin{bmatrix} x & y & 1 \end{bmatrix}^T$, where x, y are the point's position in the coordinate system of the respective digital camera's matrix.
- Calibration matrices of the digital cameras [21,26]: K_m.
- Rotation matrices of the digital cameras [25], describing their orientation in the test site coordinate space relative to the coordinate axes, determined based on data from Table 1 (row 3): R_m.
- Translation matrices of the digital cameras [25], describing their position in the test site coordinate space: $T_m = \begin{bmatrix} Xc_m & Yc_m & Zc_m \end{bmatrix}^T$, where Xc_m, Yc_m, Zc_m are the camera positions in the test site coordinate system, determined based on data from Table 1.
- Index of the initial main camera: $mcam$. This is the camera for which, at the current algorithm iteration, the deviations between adjacent jet trajectory points along the axis perpendicular to the corresponding coordinate plane are minimal. In the diagram presented in Figure A2, the main camera at the initial section of the flow is camera XY, so for this scenario, the corresponding index is chosen as $mcam$.
- Minimum step through the trajectory image point arrays: $MinStep$.
- Minimum allowable distance between adjacent trajectory points in the test site coordinate system: ε.

Output Data:

- Array of jet trajectory points in the test site coordinate system: P.

Step 1. Algorithm initialization:

1.1. Add the point with coordinates $P_N = \begin{bmatrix} 0 & 0 & 0 \end{bmatrix}^T$ ($N = 0$) in the test site system to the trajectory array P.

1.2. Select the last trajectory point $cP = P_N$ as the current point.

1.3. Select camera mcam as the main camera: $main = mcam$.

1.4. Calculate matrices used for computations: $Q_m = K_m \cdot R_m$ and $q_m = -Q_m \cdot T_m, m \in Cam$.

Set the current step through the trajectory image point arrays to the minimum: $step = MinStep$.

Step 2. Perform a forward perspective transformation [25] for the current point P with the parameters of the main camera ($m = main$):

$$cp_m = Q_m \cdot cP + q_m, cp_m = cp_m / cp_{m,2} \tag{A8}$$

Step 3. Determine the index of the point in the trajectory image from the main camera ($m = main$) closest to cp_m:

$$k = k : |p_k - cp_m| = \min_{p_i \in tp_m} (|p_i - cp_m|), \tag{A9}$$

where $| \ |$ is the vector norm.

If $k + step \geq |tp_m|$ (where $| \ |$ denotes the number of elements in the array), then end the algorithm.

Step 4. Select the point following the closest one in the array ($m = main$):

$$nextp_m = tp_{m,k+step} \tag{A10}$$

Perform an inverse perspective transformation [25] of the point $nextp_m$:

$$nextP_m = Q_m^{-1} \cdot (\lambda \cdot nextp_m - q_m), \tag{A11}$$

where $\lambda = \frac{cP_{\mathrm{TC}(m)} + [Q_m^{-1} \cdot q_m]_{\mathrm{TC}(m)}}{[Q_m^{-1} \cdot nextp_m]_{\mathrm{TC}(m)}}$, and $\mathrm{TC}(m) = \begin{cases} Z, m = XY \\ X, m = YZ \\ Y, m = XZ \end{cases}$ is a function establishing the correspondence between cameras and the coordinate axes of the test site system that are orthogonal to the planes of their matrices.

Step 5. For cameras that are not the main ($m \neq main, m \in Cam$), determine the points in the trajectory images closest to the respective forward perspective transformations of $cP = nextP_{main}$ using Formulas (A8)–(A10) (for $step = 0$), as well as their inverse perspective transformations (A11).

Step 6. Based on the points $nextP_m$ obtained from all cameras in Steps 2–5, determine the new current point cP by averaging

$$cP = \frac{\sum_{m \in C} nextP_m}{|Cam|}, \tag{A12}$$

where $|Cam|$ is the number of elements in the array Cam.

Step 7. If the distance between cP and the last trajectory point P_N is less than ε,

$$|cP - P_N| < \varepsilon, \tag{A13}$$

then $step = step + 1$, $cP = P_N$ and go to Step 2. Otherwise, if $step >$ MinStep, then $step = step - 1$.

Step 8. The current point cP is added to the trajectory array P: $N = N + 1$, $P_N = cP$. Remove points lying between the jet outlet and $nextp_m$ inclusive from the arrays tp_m ($tp_{m,i} : i \leq k, nextp_m = tp_{m,k}, m \in Cam$).

Step 9. Determine the new main camera based on the minimal deviation of coordinates in the test site system between the two last trajectory points:

$$main = \text{invTC}(ax), \tag{A14}$$

where

$$ax = ax : |P_{N,ax} - P_{N-1,ax}| = \min_{i=\langle X,Y,Z\rangle}(|P_{N,i} - P_{N-1,i}|), \tag{A15}$$

$\text{invTC}(ax) = \begin{cases} YZ, ax = X \\ XZ, ax = Y \\ XY, ax = Z \end{cases}$ is a function establishing the correspondence between the coordinate axes of the test site and the cameras whose matrices lie in planes orthogonal to them.

Go to Step 2.

The removal of points from the arrays tp_m lying between the jet outlet and $nextp_m$ in Step 8 is used to prevent incorrect algorithm behavior near the jet's height maximum when the distances between the images of the ascending and descending flows of the fire-extinguishing agent are small. This can occur, for example, in the YZ projection when crosswind speed is low, or in the XY projection at large values of the output nozzle elevation angle. In such cases, at Steps 3 and 5 of the algorithm, points lying on the adjacent section of the trajectory (ascending instead of descending) might be selected as the closest to cp_m, which would lead to error accumulation in the constructed model during subsequent iterations. Therefore, to prevent this effect, the corresponding correction is performed in Step 8.

Appendix C

Algorithm for Constructing the Line of Maximum Flow Velocities

The algorithm below implements the proposed method for constructing the fire-extinguishing agent jet trajectory from numerical simulation results. The following axes of the three-dimensional space are adopted in the algorithm description:

- OX axis: Horizontal axis coinciding with the corresponding projection of the initial flow direction.
- OY axis: Vertical axis.
- OZ axis: Horizontal axis orthogonal to the OX and OY axes.
- Coordinate origin: Located at the geometric center of the flow outlet region (jet starting point).

Input Data:

- Array of grid node coordinates $P = \langle X_i, Y_i, Z_i; i = 1 \ldots n\rangle$, where n is the number of computational grid nodes.
- Array of velocity vectors at grid nodes $V = \langle Vx_i, Vy_i, Vz_i; i = 1 \ldots n\rangle$.
- Set of nodes forming the outlet region of the investigated flow: *outlet*.
- Step length: *step*.
- Section width: *width*.
- Threshold value determining proximity of velocity to zero: ε.

Step 1. The initial point with the maximum velocity magnitude within the flow outlet region is selected and added to the flow line array: $TP_1 = P_k : |V_k| = \max_{i=1\ldots n, P_i \in oulet}(|V_i|)$ (Figure A3), where $|V_i| = \sqrt{Vx_i^2 + Vy_i^2 + Vz_i^2}$ denotes the vector magnitude.

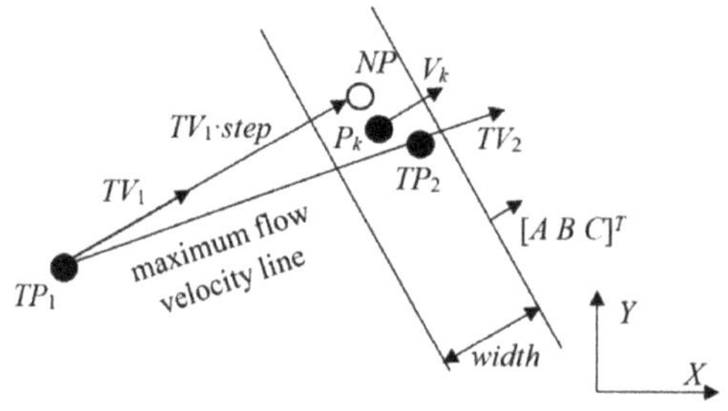

Figure A3. Explanatory diagram for the algorithm for constructing the line of maximum flow velocities.

The velocity vector with the maximum magnitude is added to the velocity array for the flow line: $TV_1 = V_k$.

The iteration counter is set: $N = 1$.

Step 2. The coordinates of the support point are calculated: $NP = TP_N + step \cdot TV_N$ (Figure A3).

Step 3. The grid node closest to the support point NP is selected: $P_k : |P_k - NP| = \min\limits_{i=1...n} (|P_i - NP|)$ (Figure A3).

Step 4. Within the vicinity of the flow section by a plane passing through point $P_k = \langle X_k, Y_k, Z_k \rangle$ and whose normal coincides in direction with the velocity V_k at that point, a search for the grid node with the maximum velocity is performed (Figure A3). For this, the parameters of the section plane equation of the form (A16) are calculated:

$$A \cdot x + B \cdot y + C \cdot z + D = 0, \tag{A16}$$

where $NV = \langle A, B, C \rangle = V_k / |V_k|$ is the plane normal,

$D = -(A \cdot X_k + B \cdot Y_k + C \cdot Z_k)$ is the free parameter of the plane equation.

Next, the distances from grid nodes to the section plane are computed using Formula (A17):

$$r(P_i) = |A \cdot X_i + B \cdot Y_i + C \cdot Z_i + D| \tag{A17}$$

From the set of grid nodes lying within the vicinity of the section plane (*width*), the point with the maximum velocity is selected and added to the flow line array (Figure A3): $TP_{N+1} = P_m : |V_m| = \max\limits_{i=1...n, r(P_i) \le width/2} (|V_i|)$.

The velocity vector with the maximum magnitude is added to the velocity array for the flow line: $TV_{N+1} = V_m$.

If $TP_{N+1} = TP_N$, then the last point in the line of maximum velocities array is removed and the algorithm stops.

If the maximum velocity magnitude is close to zero $|TV_{N+1}| \le \varepsilon$, then the end of the flow trajectory has been reached, and the algorithm stops. Otherwise, the iteration counter is incremented $N = N + 1$ and the algorithm proceeds to Step 2.

When using grids with large distances between nodes, to improve accuracy at Step 3, one can use points whose coordinates and velocity vector are calculated based on the vertices of the cell containing NP, using formulas (A18):

$$TP = \frac{\sum\limits_{P_i \in F} (VF_i \cdot P_i)}{\sum\limits_{P_i \in F} VF_i}, TV = \frac{\sum\limits_{P_i \in F} (VF_i \cdot V_i)}{\sum\limits_{P_i \in F} VF_i}, \tag{A18}$$

where F is the computational grid cell containing the support point NP,

$VF = \langle VF_i; i = 1 \ldots n \rangle$ is the array of liquid (phase) volume fraction values at the grid nodes.

References

1. Wang, M.; Chen, X.; Huang, X. Robotic firefighting: A review and future perspective. In *Intelligent Building Fire Safety and Smart Firefighting*; Springer Nature: London, UK, 2024; pp. 475–499.
2. Pozharkova, I.; Abdul-Zahra, D.S.; Ngo-Hoang, D.L.; Shavkatov, N.; Mohammed, K.A.; Salem, K.H. Evaluating the effectiveness of fire robots for the protection of transport infrastructure facilities. *Transp. Res. Procedia* **2023**, *68*, 499–504. [CrossRef]
3. Loboichenko, V.; Wilk-Jakubowski, G.; Pawlik, L.; Wilk-Jakubowski, J.L.; Shevchenko, R.; Shevchenko, O.; Rushchak, I. Review of advances in fire extinguishing based on computer vision applications: Methods, challenges, and future directions. *Sensors* **2025**, *25*, 6399. [CrossRef]
4. Raj, R.; Kos, A. A comprehensive study of mobile robot: History, developments, applications, and future research perspectives. *Appl. Sci.* **2022**, *12*, 6951. [CrossRef]
5. Lattimer, B.Y.; Huang, X.; Delichatsios, M.A.; Levendis, Y.A.; Kochersberger, K.; Manzello, S.; Suzuki, S. Use of unmanned aerial systems in outdoor firefighting. *Fire Technol.* **2023**, *59*, 2961–2988. [CrossRef]
6. Kuznetsov, G.; Kopylov, N.; Sushkina, E.; Zhdanova, A. Adaptation of fire-fighting systems to localization of fires in the premises. *Energies* **2022**, *15*, 522. [CrossRef]
7. Dhiman, A.; Shah, N.; Adhikari, P.; Kumbhar, S.; Dhanjal, I.S.; Mehendale, N. Firefighting robot with deep learning and machine vision. *Neural Comput. Appl.* **2022**, *34*, 2831–2839. [CrossRef]
8. Eslam, B.; Samir, A.; Osama, M.; Moustafa, M.; Ali, M. The evolution of firefighting robots: Bridging technology and safety: A comprehensive review. *Adv. Sci. Technol. J.* **2025**, *2*, 1–15. [CrossRef]
9. Pozharkova, I.N.; Tsarichenko, S.G.; Nemchinov, S.G. Modeling the trajectory of a fire extinguishing agent jet from a fire monitor under disturbing influences. *Bezop. Tr. V Promyshlennosti* **2022**, *11*, 7–13. [CrossRef]
10. Zhu, J.; Li, W.; Lin, D.; Zhao, G. Study on water jet trajectory model of fire monitor based on simulation and experiment. *Fire Technol.* **2019**, *55*, 773–787. [CrossRef]
11. Hou, X.; Cao, Y.; Mao, W.; Wang, Z.; Yuan, J. Models for predicting the jet trajectory and intensity drop point of fire monitors. *Fluid Dyn. Mater. Process.* **2021**, *17*, 859–869. [CrossRef]
12. Karami, A.; Menna, F.; Remondino, F. Combining photogrammetry and photometric stereo to achieve precise and complete 3D reconstruction. *Sensors* **2022**, *22*, 8172. [CrossRef]
13. Wierzbicki, D. Multi-camera imaging system for UAV photogrammetry. *Sensors* **2018**, *18*, 2433. [CrossRef]
14. Sharma, M.; Raghavendra, S.; Agrawal, S. Development of an open-source tool for UAV photogrammetric data processing. *J. Indian Soc. Remote Sens.* **2021**, *49*, 659–664. [CrossRef]
15. Tian, Y.; Ding, C.; Lin, Y.F.; Ma, S.; Li, L. Automatic feature type selection in digital photogrammetry of piping. *Comput.-Aided Civ. Infrastruct. Eng.* **2022**, *37*, 1335–1348. [CrossRef]
16. Aguera-Vega, F.; Aguera-Puntas, M.; Aguera-Vega, J.; Martinez-Carricondo, P.; Carvajal-Ramirez, F. Multi-sensor imagery rectification and registration for herbicide testing. *Measurement* **2021**, *175*, 109049. [CrossRef]
17. Cheng, H.; Zhu, J.; Wang, S.; Yan, K.; Wang, H. Firefighting water jet trajectory detection from unmanned aerial vehicle imagery using learnable prompt vectors. *Sensors* **2024**, *24*, 3553. [CrossRef]
18. Zhu, J.; Li, W.; Lin, D.; Zhao, G. Real-time monitoring of jet trajectory during jetting based on near-field computer vision. *Sensors* **2019**, *19*, 690. [CrossRef]
19. Zhu, J.; Pan, L.; Zhao, G. An improved near-field computer vision for jet trajectory falling position prediction of intelligent fire robot. *Sensors* **2020**, *20*, 7029. [CrossRef] [PubMed]
20. Zhang, C.; Zhang, R.; Dai, Z.; He, B.; Yao, Y. Prediction model for the water jet falling point in fire extinguishing based on a GA-BP neural network. *PLoS ONE* **2019**, *14*, e0221729. [CrossRef] [PubMed]
21. Elfverson, D.; Lejon, C. Use and scalability of OpenFOAM for wind fields and pollution dispersion with building-and ground-resolving topography. *Atmosphere* **2021**, *12*, 1124. [CrossRef]
22. Masoumi-Verki, S.; Haghighat, F.; Eicker, U. A review of advances towards efficient reduced-order models (ROM) for predicting urban airflow and pollutant dispersion. *Build. Environ.* **2022**, *216*, 108966. [CrossRef]
23. Ye, C.C.; Zhang, P.J.Y.; Wan, Z.H.; Yan, R.; Sun, D.J. Accelerating CFD simulation with high order finite difference method on curvilinear coordinates for modern GPU clusters. *Adv. Aerodyn.* **2022**, *4*, 7. [CrossRef]
24. Mao, R.; Lan, Y.; Liang, L.; Yu, T.; Mu, M.; Leng, W.; Long, Z. Rapid CFD prediction based on machine learning surrogate model in built environment: A review. *Fluids* **2025**, *10*, 193. [CrossRef]
25. Jain, P.; Choudhury, A.; Dutta, P.; Kalita, K.; Barsocchi, P. Random forest regression-based machine learning model for accurate estimation of fluid flow in curved pipes. *Processes* **2021**, *9*, 2095. [CrossRef]

26. Joseph, V.R.; Vakayil, A. SPlit: An optimal method for data splitting. *Technometrics* **2022**, *64*, 166–176. [CrossRef]
27. Weaver, D.S.; Mišković, S. A study of RANS turbulence models in fully turbulent jets: A perspective for CFD-DEM simulations. *Fluids* **2021**, *6*, 271. [CrossRef]
28. Kumar, S.; Bhatnagar, V. A review of regression models in machine learning. *J. Intell. Syst. Comput.* **2022**, *3*, 40–47. [CrossRef]
29. Korkmaz, K.B.; Werner, S.; Bensow, R. Verification and validation of CFD based form factors as a combined CFD/EFD method. *J. Mar. Sci. Eng.* **2021**, *9*, 75. [CrossRef]
30. Pronzato, L.; Rendas, M.J. Validation of machine learning prediction models. *New Engl. J. Stat. Data Sci.* **2023**, *1*, 394–414. [CrossRef]
31. Wu, A.; Xiao, H.; Zeng, F. A camera calibration method based on OpenCV. In Proceedings of the 4th International Conference on Intelligent Information Processing—ICIIP '19, Guilin, China, 15–17 November 2019; Association for Computing Machinery: New York, NY, USA; pp. 320–324.
32. Gupta, S.; Saluja, K.; Goyal, A.; Vajpayee, A.; Tiwari, V. Comparing the performance of machine learning algorithms using estimated accuracy. *Meas. Sens.* **2022**, *24*, 100432. [CrossRef]
33. Reinhard, E.; Stark, M.; Shirley, P.; Ferwerda, J. Photographic tone reproduction for digital images. *Semin. Graph. Pap. Push. Boundaries* **2023**, *2*, 661–670.
34. Pierson, C.; Cauwerts, C.; Bodart, M.; Wienold, J. Tutorial: Luminance maps for daylighting studies from high dynamic range photography. *Leukos* **2021**, *17*, 140–169. [CrossRef]
35. Zhang, J.; Zhang, J.; Chen, B.; Gao, J.; Ji, S.; Zhang, X.; Wang, Z. A perspective transformation method based on computer vision. In Proceedings of the 2020 IEEE International Conference on Artificial Intelligence and Computer Applications (ICAICA), Dalian, China, 27–29 June 2020; pp. 765–768.
36. Guo, K.; Ye, H.; Gu, J.; Chen, H. A novel method for intrinsic and extrinsic parameters estimation by solving perspective-three-point problem with known camera position. *Appl. Sci.* **2021**, *11*, 6014. [CrossRef]
37. Zhao, P.; Xu, J.; Ge, W.; Wang, J. A CFD-DEM-IBM method for Cartesian grid simulation of gas-solid flow in complex geometries. *Chem. Eng. J.* **2020**, *389*, 124343. [CrossRef]
38. Khosravi, M.; Javan, M. Three-dimensional flow structure and mixing of the side thermal buoyant jet discharge in cross-flow. *Acta Mech.* **2020**, *231*, 3729–3753. [CrossRef]
39. Liu, Z.; Li, Z.; Liu, J.; Wu, J.; Yu, Y.; Ding, J. Numerical study on primary breakup of disturbed liquid jet sprays using a VOF model and LES method. *Processes* **2022**, *10*, 1148. [CrossRef]
40. Li, S.J.; Zhu, L.T.; Zhang, X.B.; Luo, Z.H. Recent advances in CFD simulations of multiphase flow processes with phase change. *Ind. Eng. Chem. Res.* **2023**, *62*, 10729–10786. [CrossRef]
41. Shynybayeva, A.; Rojas-Solórzano, L.R. Eulerian–Eulerian modeling of multiphase flow in horizontal annuli: Current limitations and challenges. *Processes* **2020**, *8*, 1426. [CrossRef]
42. Octau, C.; Dbouk, T.; Watremez, M.; Méresse, D.; Lippert, M.; Schiffler, J.; Dubar, L. Liquid–solid two-phase jet in a turbulent crossflow: Experiments and simulations. *Chem. Eng. Res. Des.* **2020**, *155*, 156–171. [CrossRef]
43. Jin, W.; Xiao, J.; Ren, H.; Li, C.; Zheng, Q.; Tong, Z. Three-dimensional simulation of impinging jet atomization of soft mist inhalers using the hybrid VOF-DPM model. *Powder Technol.* **2022**, *407*, 117622. [CrossRef]
44. Garoosi, F.; Mahdi, T.F. New benchmark problems for validation and verification of incompressible multi-fluid flows based on the improved Volume-Of-Fluid (VOF) method. *Colloids Surf. A Physicochem. Eng. Asp.* **2022**, *648*, 129313. [CrossRef]
45. Höhne, T.; Porombka, P.; Moya Sáez, S. Validation of AIAD sub-models for advanced numerical modelling of horizontal two-phase flows. *Fluids* **2020**, *5*, 102. [CrossRef]
46. Setoodeh, H.; Shabestary, A.M.; Ding, W.; Lucas, D.; Hampel, U. CFD-modelling of boiling in a heated pipe including flow pattern transition. *Appl. Therm. Eng.* **2022**, *204*, 117962. [CrossRef]
47. Xiao, J.; Wu, Q.; Chen, L.; Ke, W.; Wu, C.; Yang, X.; Jiang, H. Assessment of different CFD modeling and solving approaches for a supersonic steam ejector simulation. *Atmosphere* **2022**, *13*, 144. [CrossRef]
48. Lewak, M.W.; Tępiński, J.; Klapsa, W. The use of the k-ω SST turbulence model for mathematical modeling of jet fire. *Saf. Fire Technol.* **2022**, *59*, 28–40. [CrossRef]
49. Pan, Y.; Geng, Z.; Yuan, H.; Zhai, S.; Huo, F. Numerical simulation and flow field analysis of porous water jet nozzle based on fluent. *Appl. Sci.* **2024**, *14*, 7075. [CrossRef]
50. Tran, M.K.; Panchal, S.; Chauhan, V.; Brahmbhatt, N.; Mevawalla, A.; Fraser, R.; Fowler, M. Python-based scikit-learn machine learning models for thermal and electrical performance prediction of high-capacity lithium-ion battery. *Int. J. Energy Res.* **2022**, *46*, 786–794. [CrossRef]
51. Luo, K.; Zhao, J.; Wang, Y.; Li, J.; Wen, J.; Liang, J.; Liao, S. Physics-informed neural networks for PDE problems: A comprehensive review. *Artif. Intell. Rev.* **2025**, *58*, 323. [CrossRef]
52. Lee, Y.; Kang, S. Dynamic ensemble of regression neural networks based on predictive uncertainty. *Comput. Ind. Eng.* **2024**, *190*, 110011. [CrossRef]

53. a Ilemobayo, J.; Durodola, O.; Alade, O.; J Awotunde, O.; T Olanrewaju, A.; Falana, O.; E Edu, O. Hyperparameter tuning in machine learning: A comprehensive review. *J. Eng. Res. Rep.* **2024**, *26*, 388–395. [CrossRef]
54. Lindauer, M.; Eggensperger, K.; Feurer, M.; Biedenkapp, A.; Deng, D.; Benjamins, C.; Hutter, F. SMAC3: A versatile Bayesian optimization package for hyperparameter optimization. *J. Mach. Learn. Res.* **2022**, *23*, 1–9.
55. Cuomo, S.; Di Cola, V.S.; Giampaolo, F.; Rozza, G.; Raissi, M.; Piccialli, F. Scientific machine learning through physics–informed neural networks: Where we are and what's next. *J. Sci. Comput.* **2022**, *92*, 88. [CrossRef]
56. Guo, C.; Sun, L.; Li, S.; Yuan, Z.; Wang, C. Physics-informed Kolmogorov–Arnold network with Chebyshev polynomials for fluid mechanics. *Phys. Fluids* **2025**, *37*, 095120. [CrossRef]
57. Somvanshi, S.; Javed, S.A.; Islam, M.M.; Pandit, D.; Das, S. A survey on Kolmogorov-Arnold network. *ACM Comput. Surv.* **2025**, *58*, 1–35. [CrossRef]
58. de Oliveira, J.P.S.; Alves, J.V.B.; Carneiro, J.N.E.; de Andrade Medronho, R.; Silva, L.F.L.R. Coupling a neural network technique with CFD simulations for predicting 2-D atmospheric dispersion analyzing wind and composition effects. *J. Loss Prev. Process Ind.* **2022**, *80*, 104930. [CrossRef]

robotics

MDPI

Article

Attention-Guided Path Planning: Learning Efficient Heuristics for Mobile Robot Navigation via Deep Neural Networks

Abderrahim Waga [1,*], Said Benhlima [2], Ali Bekri [2], Fatima Zahrae Saber [3], Jawad Abdouni [4], Toufik Mzili [5] and Ahmed Regragui [2]

[1] School of Digital Engineering and Artificial Intelligence, Euromed University, Fez 30000, Morocco
[2] Computer Science, Faculty of Sciences, Moulay Ismail University, Zitoun, Meknes 50000, Morocco; s.benhlima@umi.ac.ma (S.B.); a.bekri@umi.ac.ma (A.B.); ah.regragui@edu.umi.ac.ma (A.R.)
[3] Computer Science, Faculty of Sciences, Ibn Tofail University, Kenitra 14000, Morocco; fatimazahrae.saber@uit.ac.ma
[4] Department of Computer Science, National School of Applied Sciences of Kenitra, Ibn Tofail University, Kenitra 14000, Morocco; j.abdouni@enim.ac.ma
[5] Computer Science, Faculty of Sciences, Chouaib Doukkali University, Eljadida 24000, Morocco; mzili.t@ucd.ac.ma
* Correspondence: a.waga@ueuromed.org

Abstract

Path planning in cluttered environments constitutes a critical challenge for mobile robotics. Although optimal solutions can be obtained by classical methods such as A*, they have the disadvantage of being computationally expensive in complex environments. In this paper, we propose a novel deep learning-based framework for 2D trajectory prediction in grid environments. The framework employs attention mechanisms specifically designed for path planning tasks. In particular, we design an Attention U-Net architecture that employs attention gates for effective path area focusing and residual connections for efficient feature selection. To validate our method, the Attention U-Net architecture is trained on 5000 randomly sampled 40×40 environments and tested on a separate test set of 200 environments. The experimental results show that the Attention U-Net architecture significantly outperforms the A* algorithm. It expands 62% fewer nodes (207.6 vs. 543.11) and achieves near-optimal path lengths (99.8% of optimal) and planning speed (0.78 ms vs. 1.19 ms). Furthermore, the Attention U-Net architecture achieves a 100% success rate for A* path planning with the attention heuristic, demonstrating the effectiveness of the attention heuristic for path planning.

Keywords: mobile robot navigation; path planning; deep learning; attention mechanisms; U-Net; A*

1. Introduction

Autonomous mobile robot navigation in cluttered environments constitutes a critical challenge that necessitates efficient path planning algorithms capable of rapidly computing feasible trajectories from a start position to a goal. This fundamental problem has been widely studied, as evidenced by extensive surveys for trajectory planning [1], path planning problems [2,3], and autonomous vehicle navigation systems [4]. Such studies reflect the continued challenges of such problems, including computational complexity, real-time execution, and adaptability. While optimal solutions for such problems have been obtained using traditional graph algorithms such as A* [5] and Dijkstra's algorithm [6], such algorithms are often computationally intensive. These methods require a significant search over the configuration space, causing computational inefficiencies, particularly in complex environments. For instance, in a 40 × 40 grid map, A* search may require the expansion of hundreds or thousands of nodes to find the shortest path. Although this issue is alleviated by the use of heuristic search, the issue remains: the need for a search over the configuration space is inherent in classical planners, making them inefficient for real-time tasks and complex environments. More recent developments in deep learning present a promising alternative [7,8]. In contrast to search methods, a proper encoding of paths may now be learned directly from the data. Deep learning methods are capable of learning a direct mapping from the environment representation to the trajectory, effectively circumventing the search process altogether [9]. This paradigm shift in planning, from search to learning, presents a wide range of exciting possibilities for real-time robotic path planning. In this paper, we present a novel learning framework for the task of predicting the trajectory for a robot in a two-dimensional grid map environment, using deep learning methods. Using the occupancy grid map, along with the start and goal locations, our goal is to learn to predict a valid path in the form of a binary mask. In contrast to classical planners, our proposed solution learns a direct mapping from the environment to the trajectory, allowing for fast inference for real-time tasks.

Figure 1 illustrates the overall workflow of our proposed attention-guided path planning framework. Specifically, it depicts the end-to-end pipeline we designed to transform the computational heavy lifting required by classical search into fast learned inference.

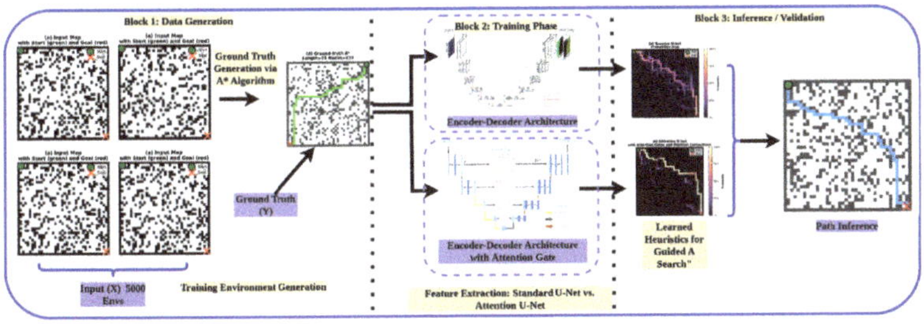

Figure 1. Overview of the proposed attention-guided path planning framework. The workflow illustrates the transformation of classical search into fast learned inference. The left side shows the generation of the synthetic dataset using A*, the middle section depicts the training of the neural network architectures, and the right side demonstrates the final hybrid planning. In the map visualizations, green elements denote start positions, red elements denote goal positions, and blue lines indicate the planned paths.

First, the system generates a synthetic dataset of random occupancy grid maps where the optimal path is computed using A*, acting as an expert demonstrator. This ground truth data is then used to train two deep neural network architectures: a standard U-Net and our proposed Attention U-Net. The models learn to predict a probability map that highlights the optimal traversal area. Finally, these predicted maps serve as learned heuristics to guide the A* search, significantly pruning the exploration space while ensuring optimal pathfinding.

A* is used as the benchmark and expert label generator for three reasons: (i) with an admissible heuristic on grid maps, it provides deterministic shortest-path labels suitable for stable supervision, (ii) compared to Dijkstra, it preserves optimality while reducing unnecessary expansions via heuristic guidance, and (iii) unlike sampling-based planners such as RRT*, it avoids stochastic label variability that can weaken the consistency of pixel-level trajectory supervision.

https://doi.org/10.3390/robotics15050096

1.1. Contributions

Our main contributions are:

1. **Novel Architecture**: We propose an Attention U-Net architecture specifically designed for path planning. This architecture combines:
   - Attention gates in skip connections for feature selection, enabling the model to focus on path-relevant regions.
   - Residual connections in convolutional blocks for improved gradient flow.
   - U-Net encoder–decoder structure with skip connections for dense prediction.
2. **Superior Performance**: We demonstrate that Attention U-Net + A* achieves superior performance compared to classical A*: it expands 62% fewer nodes (207.6 vs. 543.11) while maintaining near-optimal path lengths (99.8% of optimal) and faster planning time (0.78 ms vs. 1.19 ms).
3. **Hybrid Planning Approach**: Instead of claiming novelty for learning-guided planning in general, we contribute a task-specific integration for 2D occupancy grids: Attention U-Net probability maps are embedded into weighted A* cost shaping. This hybrid design provides a practical efficiency–optimality trade-off with consistently high success rates and near-optimal paths.

1.2. Paper Organization

The rest of this paper is organized as follows: Section 2 presents related work on path planning and deep learning for robotics. Section 3 describes our approach to path planning using deep learning for robotics. Section 4 describes the experimental setting and outcomes. Section 5 discusses the implications of this work. Section 6 presents concluding remarks on this work and suggestions for further research.

2. Related Work

In this section, we provide a detailed literature review of path planning techniques, both classical and machine learning-based. We discuss the pros and cons of both types of techniques and identify the research gap filled by our study.

2.1. Classical Path Planning

In this context, Kabir et al. [10] have proposed a significant improvement in the A* search strategy by introducing a Robot Motion Block, accompanied by an adaptive cost function, which reduces the number of explored nodes by nearly 94%, thus significantly increasing the rate of convergence compared to traditional A*. At the same time, in order to effectively handle the kinematic constraints associated with robotic motion, a bidirectional A* search strategy has been proposed by Yin et al. [11], which incorporates the scale factor as well as the angle factor in the cost function. Furthermore, in order to eliminate the angular properties associated with the motion planning in a grid-based environment, a cubic B-spline curve smoothing strategy has been integrated with the proposed bidirectional A*. The importance of such a strategy has been highlighted in a previous study by Duchon et al. [12], which discussed the importance of "any-angle" search algorithms, such as Theta* or Phi*, in order to effectively handle the geometric limitations associated with a traditional grid-based motion planning strategy. Considering the limitations associated with the use of deterministic motion planning algorithms in high-dimensional spaces, RRT* has gained significant popularity as a motion planning strategy, as discussed in the literature. However, the limitations associated with the slow rate of convergence have forced researchers to propose directional biases in motion planning algorithms.

https://doi.org/10.3390/robotics15050096

Fan et al. [13] suggested the hybrid approach combining rapidly exploring random trees (RRT*) with the artificial potential field (APF) approach for the path planning problem of the UAV. Through the application of the goal-biased approach, the blindness caused by the random approach can be well resolved. With a similar aim of improving environmental adaptability, the DBVSB-P-RRT* algorithm was proposed by Guo et al. [14] and was shown to reduce planning time by more than 40%.

Other researchers have focused on the intrinsic quality of the search tree. Dong et al. [15] proposed the FHQ-RRT* algorithm, which is motivated by dynamic sparse sampling and the generation of "KeyPoints" to optimize the selection of the parents, while Zhu et al. [16] proposed the dual-weighted RRT* algorithm (GAO-RRT*), which directs the search tree towards the goal as well as the proximate obstacles to reduce the path length, contrary to the traditional belief in the advantage of the broader obstacle avoidance.

More recently, the trend with the greatest promise is related to the combination of traditional planners and techniques from artificial intelligence and deep learning. The latter techniques frequently try to address local minimum problems, common in reactive approaches. For example, Zhang et al. [17] proposed a system that integrates A*, the dynamic window approach (DWA), and deep Q-networks (DQNs). In the proposed system, the agent is responsible for making decisions on how the robot can change trajectories in real-time. In other words, the agent is responsible for escaping dead ends, where reactive techniques may fail. In another example, Song et al. [18] proposed an approach related to the integration of obstacle prediction techniques. The authors proposed an approach that integrates a Random Forest classifier, an S-RRT algorithm, and Ant Colony Optimization. The proposed approach demonstrates how supervised learning can be employed to significantly enhance local obstacle avoidance techniques. For 3D environments, another approach related to the combination of enhanced potential fields and B-spline curves has been proposed by Liu et al. [19]. Recent studies also highlight complementary planning directions that are closely related to the present motivation. Menon et al. propose a homotopy-aware, efficiently adaptive state-lattice strategy for cluttered mobile-robot navigation, where the key idea is to preserve topological diversity of feasible route classes while keeping search tractable [20]. Li et al. introduce an obstacle-avoiding manipulator planner based on recursive segmentation point migration combined with progressive inverse kinematics, aiming to improve collision-safe convergence in constrained workspaces [21]. Zhang et al. present a cosine-similarity-based multi-target planning method for cable-driven manipulators, focusing on efficient target ordering and reduced overall traversal cost in multi-goal tasks [22]. Hu et al. develop an efficient RRT-based kinodynamic framework for wheeled robots that jointly targets motion feasibility, smoothness, and path compactness under dynamic constraints [23]. Collectively, these studies reinforce the value of integrating structural priors, optimization mechanisms, and computational efficiency in modern robotic planning systems.

2.2. Machine Learning-Based Approaches

While classical and sampling-based algorithms rely on explicit geometric calculations, the recent paradigm shift toward machine learning (ML) enables robots to learn navigation policies directly from environmental data or expert demonstrations. This change can be best illustrated with regard to the domain of Deep Reinforcement Learning (DRL), wherein agents are able to achieve the best strategies via trial-and-error-based learning. This can be best explained with regard to the initial work done by Lei et al. [24] on the application of the Double Deep Q-Network (DDQN) with regard to the development of dynamic path planning in environments that are not well known. This approach was able to mitigate the instability that often comes with the application of the traditional Q-learning approach,

https://doi.org/10.3390/robotics15050096

resulting in the development of the path planning approach without the need for any initial maps. Building on the initial work done on the application of the DDQN approach, Zhang et al. [25] were able to address the problems that often come with the application of the initial approach with regard to the development of the path planning approach via the application of the GAP-SAC approach. This approach was able to allow the robots to better attend to the environmental features with the application of the gated attention mechanism, resulting in better performance compared to the application of traditional algorithms like the TD3. Moreover, the work done by Jaramillo-Martínez et al. [26] emphasized the need for the application of the reward shaping approach with regard to the application of the reinforcement learning approach. This was done via the development of the reward function strategy that can help in minimizing not only the distance but also the turns, resulting in the achievement of a 50% reduction in turns with regard to the application of the classical A* approach. Parallel to RL, Supervised Learning and Imitation Learning have been utilized to mimic expert behaviors while reducing computational overhead. Chehelgami et al. [27] introduced a safe global path planning method based on Recurrent Neural Networks (RNNs). They designed a novel loss function that not only learns the path generation but also enforces an adjustable safety margin around obstacles, allowing for the rapid generation of collision-free paths in constant time. Similarly, Waga et al. [28] explored learning from demonstration (LfD) to transfer the optimality of the A* algorithm to a machine learning model. By employing ensemble methods to learn from an "expert" A* planner, they enabled mobile robots to acquire robust navigation policies that mimic optimal behaviors without the heavy runtime search cost associated with classical planners. Most recently, research has expanded into advanced perception and large language models (LLMs) to handle high-level reasoning and complex unstructured environments. In agricultural robotics, Diao et al. [29] proposed a navigation line extraction algorithm based on an improved YOLOv8s network. By combining deep learning-based visual detection with the least squares method, they achieved high adaptability in extracting navigation paths for corn-spraying robots in complex field environments. Pushing the boundaries of cognitive navigation, Waga et al. [30] introduced a hierarchical framework leveraging GPT-4o mini. Their approach utilizes the LLM as a high-level planner to set general waypoints based on contextual reasoning, while an ensemble learning model handles low-level action execution. This hybrid structure combines the semantic understanding of large language models with the precise control of traditional learning methods, marking a significant step toward intelligent, reasoning-capable autonomous navigation.

2.3. Research Gap

Although path planning has been extensively studied, there remain critical gaps that prevent it from being applied well in reality. Based on a review of classical path planning algorithms [10–19] and learning-based path planners [24–30], there are three major gaps: (1) classical path planners have computational costs that limit their application to real-time situations, (2) current learning-based path planners do not incorporate attention mechanisms specific to path planning tasks, and (3) hybrid approaches have not been able to achieve both efficiency and near-optimality. Our proposed Attention U-Net model remedies all three gaps by incorporating attention gates, probability maps learned to guide A* searches optimally, and a 62% reduction in node expansions with near-optimal paths. Table 1 summarizes the gaps and our contributions.

https://doi.org/10.3390/robotics15050096

Table 1. Summary of identified research gaps and contributions.

Aspect	Gap in Existing Work	Our Contribution
Computational Efficiency	Classical planners (A*, Dijkstra, RRT*) expand many nodes in cluttered environments, leading to high computation time and limited real-time applicability.	Attention U-Net + A* reduces node expansions by 62% compared to classical A* while maintaining near-optimal path length, enabling faster planning suitable for real-time use.
Model Architecture	Many learning-based methods lack architectures tailored to path planning; they treat all regions equally and do not exploit attention to focus on path-relevant areas.	We introduce an Attention U-Net with attention gates in skip connections, allowing the model to focus on path-relevant regions and suppress irrelevant background features.
Hybrid Integration	Existing hybrid methods often require complex integration and still rely heavily on exhaustive search, failing to fully leverage learned heuristics for efficiency gains.	Our model produces probability maps that directly serve as effective heuristics for A*, combining learned guidance with classical optimality guarantees in a simple, practical pipeline.

3. Proposed Methodology

In this section, we outline our approach to the learning process in path planning. This begins with a formal definition of the path planning problem and the formulation of the mathematical framework. Next, we introduce the baseline architecture, followed by the proposed architecture incorporating the attention mechanism. Finally, we define the loss functions used for the path planning tasks and the training process used to train the model to learn the heuristics.

3.1. Problem Formulation

We consider the problem of predicting robot trajectories in 2D grid environments (Figure 2). The input consists of the following:

- **Occupancy Grid Map** $G \in \{0,1\}^{H \times W}$: Binary grid where 1 represents free space, and 0 represents obstacles or boundaries. In our experiments, $H = W = 40$.
- **Start Position** $s = (s_x, s_y)$: Initial robot position, where $s_x, s_y \in \{0, \ldots, H-1\} \times \{0, \ldots, W-1\}$ and $G[s_x, s_y] = 1$ (start must be in free space).
- **Goal Position** $g = (g_x, g_y)$: Target position, with the same constraints as the start position.

Figure 2. Illustrations of binary occupancy grid maps $G \in \{0,1\}^{H \times W}$ representing 2D environments with robot start and goal locations.

The output is a **trajectory mask** $T \in \{0,1\}^{H \times W}$ where $T[i,j] = 1$ indicates that the cell (i,j) is part of the predicted path. Ideally, the predicted path should

https://doi.org/10.3390/robotics15050096

1. Connect the start s to the goal g (connectivity constraint);
2. Avoid obstacles: $T[i,j] = 1 \implies G[i,j] = 1$ (feasibility constraint);
3. Minimize path length (optimality objective).

We encode the input as a 3-channel tensor $X \in \mathbb{R}^{3\times H\times W}$:

- Channel 0: Occupancy grid G (normalized to $[0,1]$).
- Channel 1: Start position mask S where $S[i,j] = 1$ if $(i,j) = s$, else 0.
- Channel 2: Goal position mask G_{mask} where $G_{mask}[i,j] = 1$ if $(i,j) = g$, else 0.

The model learns a mapping $f_\theta : \mathbb{R}^{3\times H\times W} \to [0,1]^{H\times W}$ that predicts the trajectory mask. The output is a probability map $P = f_\theta(X)$ where $P[i,j] \in [0,1]$ represents the probability that cell (i,j) is part of the path. We threshold this probability map to obtain the binary trajectory mask: $T[i,j] = \mathbf{1}[P[i,j] > \tau]$ where $\tau = 0.5$ is a threshold parameter.

3.2. Baseline Architectures

To provide context for our proposed attention-based U-Net, we start by describing a conventional U-Net model, which serves as the reference model for comparison with our proposal. In addition, a complete understanding of the reference model helps us evaluate the improvements offered by attention mechanisms. In this section, we describe a conventional U-Net model and its modification for solving path planning problems, as detailed in Algorithm 1.

Algorithm 1 U-Net forward pass for path prediction.

Require: Input map $X \in \mathbb{R}^{3\times 40\times 40}$ (occupancy + start + goal)
Ensure: Predicted trajectory mask $P \in [0,1]^{40\times 40}$
Encoder:
for block $i = 1$ to 3 **do**
  $x \leftarrow \text{ConvBlock}_i(x)$ {Two conv layers + BN + ReLU}
  if $i < 3$ **then**
    $x \leftarrow \text{MaxPool2d}(x)$ {Downsample by 2}
    Store x in skip connection e_i
  end if
end for
$x \leftarrow \text{Bottleneck}(x)$ {Convolutional block at 5 × 5}
Decoder:
for block $i = 3$ down to 1 **do**
  $x \leftarrow \text{TransposeConv}(x)$ {Upsample by 2}
  $x \leftarrow \text{Concat}(x, e_i)$ {Skip connection}
  $x \leftarrow \text{ConvBlock}_i(x)$
end for
$P \leftarrow \text{Sigmoid}(\text{Conv1} \times 1(x))$ {Output layer}
return P

The selection of U-Net as the backbone is motivated by three practical considerations in this study. First, trajectory prediction in occupancy grids is a dense pixel-level task, where preserving fine spatial details (obstacle boundaries, narrow passages, start/goal anchors) is essential; U-Net skip connections are well suited to this requirement. Second, this work targets controlled architectural attribution: by keeping the encoder–decoder family fixed and introducing attention/residual modifications incrementally, the measured gain can be directly attributed to the proposed modules rather than to a wholesale backbone change. Third, under the current CPU-oriented experimental setting and training budget, U-Net provides a stable and computationally tractable baseline for fair comparison across variants.

https://doi.org/10.3390/robotics15050096

U-Net Baseline

We use a standard U-Net architecture [31] as our deterministic baseline. The U-Net consists of a symmetric encoder–decoder structure with skip connections.

Encoder: The encoder consists of three downsampling blocks. Each block contains:

- Two convolutional layers with a kernel size of 3 × 3 and padding of 1.
- Batch normalization and ReLU activation.
- Max pooling with a stride of 2 for downsampling (except in the last block).

The encoder progressively reduces spatial resolution: $40 \times 40 \rightarrow 20 \times 20 \rightarrow 10 \times 10 \rightarrow 5 \times 5$, while increasing channel depth: $3 \rightarrow 32 \rightarrow 64 \rightarrow 128$.

Bottleneck: This is a convolutional block that operates at the lowest resolution, which is 5 × 5, with 256 channels.

Decoder: The decoding stage makes use of transposed convolutions (stride of 2) for upsampling, followed by convolutional blocks. Skip connections connect the features of the encoder to those of the decoder at equivalent resolutions.

Output Layer: The channel dimension is reduced to 1 using a convolution, followed by a sigmoid activation to produce probability values ranging from 0 to 1.

The U-Net architecture is especially suited to dense predictions, such as path planning, as the skip connections provide the ability to combine the low-level spatial details (e.g., boundary of obstacles, start/goal points) with the high-level semantic details (e.g., path and connectivity).

3.3. *Proposed Attention U-Net*

The proposed architecture extends the original U-Net model with attention mechanisms that are carefully designed to improve path planning performance. The key novelty of our approach is the introduction of attention gates in the skip connections, which help the model focus on path information and ignore irrelevant background information. This subsection outlines the architectural design for better performance than the original model, as detailed in Algorithm 2.

We enhance the basic U-Net structure by adding attention gates [32] to the skip connections. The important part is that attention gates enable the decoder to focus on relevant areas of the trajectory using attention coefficients that weight the features of the encoder. An attention gate takes two inputs:

- **Gate signal** $g \in \mathbb{R}^{B \times C_g \times H_g \times W_g}$: features from the decoder (upsampled, coarser resolution), where B is the batch size, C_g is the number of gate channels, and (H_g, W_g) is the spatial resolution.
- **Skip signal** $x \in \mathbb{R}^{B \times C_x \times H_x \times W_x}$: features from the encoder (finer resolution), where C_x is the number of skip channels and (H_x, W_x) is the spatial resolution.

The attention coefficient $\alpha \in [0, 1]^{B \times 1 \times H_x \times W_x}$ is computed through the following steps:

1. **Feature projection**: Both signals are projected to a common intermediate dimension $C_{int} = C_x/2$ using 1 × 1 convolutions:

$$g' = W_g(g) \in \mathbb{R}^{B \times C_{int} \times H_g \times W_g} \tag{1}$$

$$x' = W_x(x) \in \mathbb{R}^{B \times C_{int} \times H_x \times W_x} \tag{2}$$

where $W_g : \mathbb{R}^{C_g} \rightarrow \mathbb{R}^{C_{int}}$ and $W_x : \mathbb{R}^{C_x} \rightarrow \mathbb{R}^{C_{int}}$ are 1 × 1 convolutional layers with batch normalization.

https://doi.org/10.3390/robotics15050096

2. **Spatial alignment**: If g' and x' have different spatial dimensions, g' is upsampled via bilinear interpolation to match x':

$$\tilde{g} = \text{Interpolate}(g', \text{size} = (H_x, W_x)) \in \mathbb{R}^{B \times C_{int} \times H_x \times W_x} \tag{3}$$

3. **Attention coefficient computation**: The aligned features are combined and passed through a ReLU activation, followed by a 1×1 convolution $\psi : \mathbb{R}^{C_{int}} \rightarrow \mathbb{R}^1$ and sigmoid activation:

$$\alpha = \sigma(\psi(\text{ReLU}(\tilde{g} + x'))) \in [0,1]^{B \times 1 \times H_x \times W_x} \tag{4}$$

The modulated skip connection is then computed via element-wise multiplication:

$$x'' = x \odot \alpha \in \mathbb{R}^{B \times C_x \times H_x \times W_x} \tag{5}$$

where $\odot$ denotes element-wise multiplication (broadcasting α across the channel dimension). This allows the model to suppress irrelevant background features and focus on path-relevant regions by adaptively weighting encoder features based on decoder context. We also incorporate residual connections within convolutional blocks to improve gradient flow during training. Each convolutional block computes:

$$\text{output} = \text{ReLU}(\text{BN}(\text{Conv}(\text{input}))) + \text{input} \tag{6}$$

This enables training deeper networks and helps preserve important features through the network.

Algorithm 2 Attention U-Net forward pass.

Require: Input map $X \in \mathbb{R}^{3 \times 40 \times 40}$
Ensure: Predicted trajectory mask $P \in [0,1]^{40 \times 40}$
Encoder:
for block $i = 1$ to 3 **do**
  $x \leftarrow \text{ConvBlockResidual}_i(x)$ {With residual connection}
  if $i < 3$ **then**
    $x \leftarrow \text{MaxPool2d}(x)$
    Store x in skip connection e_i
  end if
end for
$x \leftarrow \text{Bottleneck}(x)$
Decoder with Attention Gates:
for block $i = 3$ down to 1 **do**
  $x \leftarrow \text{TransposeConv}(x)$ {Upsample decoder features}
  $\alpha_i \leftarrow \text{AttentionGate}(x, e_i)$ {Compute attention coefficients}
  $e'_i \leftarrow e_i \odot \alpha_i$ {Modulate skip connection}
  $x \leftarrow \text{Concat}(x, e'_i)$
  $x \leftarrow \text{ConvBlockResidual}_i(x)$
end for
$P \leftarrow \text{Sigmoid}(\text{Conv}1 \times 1(x))$
return P

3.4. Loss Functions

The choice of loss function has a large impact on the quality of predictions of the path. Path planning introduces several unique challenges for loss-function optimization, including class imbalance, because path pixels are a minority, and path continuity. This section describes loss functions to handle these problems, from general loss functions to specially designed loss functions to offer guidance consistent with an A* heuristic.

https://doi.org/10.3390/robotics15050096

3.4.1. Standard Losses

We use several loss functions tailored for path planning tasks, each addressing different aspects of the problem. The binary cross-entropy (BCE) [33] loss serves as the standard pixel-wise classification loss, defined as follows:

$$\mathcal{L}_{BCE} = -\frac{1}{HW}\sum_{i,j}[T[i,j]\log P[i,j] + (1 - T[i,j])\log(1 - P[i,j])] \tag{7}$$

However, BCE suffers from class imbalance: path pixels are typically less than 5% of the image, causing the model to predict all zeros. To address this limitation, we use Dice loss, which handles class imbalance by focusing on the overlap between predicted and target masks. The Dice loss is defined as follows:

$$\mathcal{L}_{Dice} = 1 - \frac{2|P \cap T| + \epsilon}{|P| + |T| + \epsilon} \tag{8}$$

where $|P \cap T|$ is the intersection, $|P|$ and $|T|$ are the sums of predicted and target masks, and $\epsilon = 1$ is a smoothing term. Additionally, we employ a combined loss that integrates multiple objectives through a weighted combination of BCE, Dice, and continuity terms:

$$\mathcal{L}_{Combined} = \alpha_{BCE}\mathcal{L}_{BCE} + \alpha_{Dice}\mathcal{L}_{Dice} + \alpha_{cont}\mathcal{L}_{Continuity} \tag{9}$$

where $\mathcal{L}_{Continuity}$ penalizes disconnected paths by measuring the number of connected components, ensuring that predicted paths maintain connectivity from start to goal.

3.4.2. A*-Optimized Loss

In the context of models used as cost heuristics in A*, we introduce a tailored loss function that addresses the unique needs of heuristic search. The basic idea is that A* performance can be enhanced by probability maps that assign high probability to optimal paths with strong peaks and low probability everywhere else to provide a strong contrast, and gradients that transition smoothly to the path rather than making binary predictions. The A*-Optimized Loss combines several components to achieve these goals. To do this, we also include focal loss [34], which focuses on the difficult examples with a high gamma value. The focal loss is given as follows:

$$\mathcal{L}_{Focal} = -\alpha(1 - P)^{\gamma}T\log P - (1 - \alpha)P^{\gamma}(1 - T)\log(1 - P) \tag{10}$$

where $\alpha = 0.25$ and $\gamma = 3.0$. The $(1 - P)^{\gamma}$ term downweights easy examples (high-confidence correct predictions) and focuses learning on hard examples, which is particularly important for learning precise path boundaries.

To encourage clear probability peaks, we employ sharpness loss, which penalizes flat distributions by measuring the entropy-like term:

$$\mathcal{L}_{Sharpness} = \frac{1}{HW}\sum_{i,j}P[i,j](1 - P[i,j]) \tag{11}$$

This loss is minimized when $P[i,j] \in \{0, 1\}$, encouraging sharp, confident predictions rather than uncertain probability distributions.

Additionally, we use path width loss to penalize overly thick predictions by measuring the average width of predicted paths:

$$\mathcal{L}_{Width} = \frac{1}{|T|}\sum_{(i,j):T[i,j]=1} \text{distance to nearest boundary} \tag{12}$$

https://doi.org/10.3390/robotics15050096

This component encourages the model to predict narrow, precise paths rather than thick regions that may include unnecessary detours.

The total A*-optimized loss integrates all these components through a weighted combination:

$$\mathcal{L}_{A*} = \lambda_{focal}\mathcal{L}_{Focal} + \lambda_{dice}\mathcal{L}_{Dice} + \lambda_{sharp}\mathcal{L}_{Sharpness} + \lambda_{width}\mathcal{L}_{Width} \tag{13}$$

with weights $\lambda_{focal} = 1.0$, $\lambda_{dice} = 0.5$, $\lambda_{sharp} = 0.3$, and $\lambda_{width} = 0.2$. These weights balance the different objectives, prioritizing focal loss for hard example learning while incorporating Dice loss for overlap maximization, sharpness for clear predictions, and width constraints for path precision, as detailed in Algorithm 3.

Algorithm 3 A*-optimized loss computation.

Require: Predicted probabilities $P \in [0,1]^{H\times W}$, target mask $T \in \{0,1\}^{H\times W}$
Ensure: Total loss $\mathcal{L}_{A*}$
1: $\mathcal{L}_{Focal} \leftarrow -\alpha(1-P)^{\gamma}T\log P - (1-\alpha)P^{\gamma}(1-T)\log(1-P)$
2: $\mathcal{L}_{Dice} \leftarrow 1 - \frac{2|P\cap T|+\epsilon}{|P|+|T|+\epsilon}$
3: $\mathcal{L}_{Sharpness} \leftarrow \frac{1}{HW}\sum_{i,j} P[i,j](1-P[i,j])$
4: $\mathcal{L}_{Width} \leftarrow \frac{1}{|T|}\sum_{(i,j):T[i,j]=1}$ distance to nearest boundary
5: $\mathcal{L}_{A*} \leftarrow \lambda_{focal}\mathcal{L}_{Focal} + \lambda_{dice}\mathcal{L}_{Dice} + \lambda_{sharp}\mathcal{L}_{Sharpness} + \lambda_{width}\mathcal{L}_{Width}$
6: **return** $\mathcal{L}_{A*}$

3.5. Training Procedure

The training process plays a crucial role in learning useful path planning heuristics. This section outlines the process of dataset generation, training setup, and model evaluation. Each part of the training process is described in depth to guarantee reproducibility. This includes dataset generation, training setup, and model evaluation.

3.5.1. Dataset Generation

In order to train the Attention U-Net model, we create a dataset of 5000 random 40×40 grid world environments. This dataset is developed in a systematic manner to ensure the diversity and quality of the training data.

Environment creation follows a multi-step procedure. To begin, an empty grid is created, and obstacles are randomly placed in cells with a given probability $p - obs = 0.25$ in inner cells, while boundary cells are set to obstacles in a deterministic fashion to form confining walls and to ensure that all possible paths stay inside the grid cells. Next, start and end points are chosen from the list of non-obstacle cells; in other words, they are often placed in opposite corners to ensure non-trivial paths requiring navigation around obstacles, but they can be placed anywhere in the free space. The ground truth path is then found using the A* algorithm and Manhattan heuristic, which returns optimal paths useful for training. The obtained path is then represented as a binary mask where path cells have a value of 1, and all other cells have a value of 0.

Each data sample in the dataset is represented as a three-channel tensor, where the channels correspond to the occupancy grid, the start position mask, and the goal position mask. The shape of the input tensor is (3, 40, 40), and the shape of the target is a binary trajectory mask (1, 40, 40). Additional information is stored for every sample, including the start and goal positions, path length, and success flag. The whole dataset is split into a training (80% and 4000 samples) and a validation (20% and 1000 samples) subset, where the validation environments are created with a different random seed to avoid any leakage between the training and validation sets. At this stage, training and test environments are procedurally generated occupancy grids and do not yet include structured floor-plan datasets (e.g., corridor–doorway layouts, office maps, or warehouse maps).

https://doi.org/10.3390/robotics15050096

3.5.2. Training Details

Training configuration plays a critical role in achieving good model performance. We use the Adam optimizer [35] with a learning rate of $\eta = 10^{-3}$, weight decay of $\lambda_{wd} = 10^{-4}$ for L2 regularization, and standard beta parameters ($\beta_1 = 0.9$, $\beta_2 = 0.999$). The batch size is set to 64 samples per batch, which provides a good balance between training stability and computational efficiency. We tested larger batch sizes (128) but found no significant improvement while requiring more memory.

Training is conducted for a maximum of 50 epochs, with the use of early stopping to prevent overfitting. Training stops automatically if, for 10 successive epochs, the validation loss does not improve (patience = 10), and the model with the lowest validation loss is saved. To better train the model, a ReduceLROnPlateau learning rate scheduler is used to decrease the learning rate by a factor of 0.5 if the validation loss has plateaued for 5 successive epochs.

In order to improve generalization and robustness, data augmentation is performed during the training procedure. This involves randomly flipping the images horizontally and vertically with a probability of 0.5 for both, and randomly rotating them by 90 degrees (choices are 0, 90, 180, and 270 degrees) with the same probability. The start and destination points are transformed based on the data augmentation.

All training is performed on a CPU. The training process is computationally intensive, with typical training time for 50 epochs taking approximately 11 h on a standard CPU. To prevent overfitting and improve generalization, we apply several regularization techniques. Dropout with probability $p_{drop} = 0.2$ is applied in convolutional blocks, batch normalization is used after each convolution to stabilize training, and weight decay provides L2 regularization where $\lambda_{wd} = 10^{-4}$.

3.5.3. Evaluation Metrics

To properly analyze the effectiveness of our Attention U-Net model, we use a variety of evaluation metrics that measure different aspects of path planning. Using these evaluation metrics, we can analyze not only the direct prediction task but also the effectiveness of our model as a heuristic for traditional planning algorithms, as detailed in Algorithm 4.

Success Rate: Percentage of environments where the predicted path connects the start to the goal. A path is considered successful if there exists a connected sequence of cells from start to goal in the predicted mask. We check connectivity using breadth-first search.

Path Length: Average length of successful paths, measured as the number of cells in the path. We compare against the optimal A* path length.

IoU (Intersection over Union): Overlap between predicted and ground-truth masks:

$$\text{IoU} = \frac{|P \cap T|}{|P \cup T|} = \frac{\sum_{i,j} \min(P[i,j], T[i,j])}{\sum_{i,j} \max(P[i,j], T[i,j])} \quad (14)$$

where P and T are predicted and target binary masks (after thresholding).

Node Expansions: For hybrid approaches (learned model + A*), the number of nodes explored by A* during planning. This measures the efficiency of the learned heuristic.

Inference Time: Wall-clock time to compute a prediction, measured in milliseconds. This includes the model forward pass and any post-processing.

https://doi.org/10.3390/robotics15050096

Algorithm 4 Model training pipeline.

Require: Dataset $\mathcal{D} = \{(X_i, T_i)\}_{i=1}^{N}$, model f_θ, loss function $\mathcal{L}$
Ensure: Trained model f_{θ^*}

Split $\mathcal{D}$ into train $\mathcal{D}_{train}$ and validation $\mathcal{D}_{val}$
Initialize model parameters θ randomly
$best\_val\_loss \leftarrow \infty$, $patience\_counter \leftarrow 0$
for epoch $= 1$ to max_epochs **do**
  Training Phase:
  for each batch $(X_b, T_b) \in \mathcal{D}_{train}$ **do**
    Apply data augmentation to (X_b, T_b)
    $P_b \leftarrow f_\theta(X_b)$
    $\mathcal{L}_{batch} \leftarrow \mathcal{L}(P_b, T_b)$
    $\theta \leftarrow \theta - \eta \nabla_\theta \mathcal{L}_{batch}$ {Adam optimizer}
  end for
  Validation Phase:
  $val\_loss \leftarrow 0$
  for each batch $(X_b, T_b) \in \mathcal{D}_{val}$ **do**
    $P_b \leftarrow f_\theta(X_b)$
    $val\_loss \leftarrow val\_loss + \mathcal{L}(P_b, T_b)$
  end for
  if $val\_loss < best\_val\_loss$ **then**
    $best\_val\_loss \leftarrow val\_loss$
    Save model checkpoint f_{θ^*}
    $patience\_counter \leftarrow 0$
  else
    $patience\_counter \leftarrow patience\_counter + 1$
    if $patience\_counter \geq patience$ **then**
      break {Early stopping}
    end if
  end if
  Update learning rate scheduler
end for
return f_{θ^*}

4. Experiments and Results

This section presents a thorough analysis of the Attention U-Net technique for path planning. The technique is compared with classical planning algorithms with regard to its performance across a set of parameters. The subsequent subsections cover the experimental environment, direct prediction results, hybrid planning results, and analysis.

4.1. Experimental Setup

We compare all models on a held-out test set of 200 procedurally generated environments. The procedurally generated environments are created in the same way as in the training data, but with a different random number seed. For each model, we measure:

1. **Direct trajectory prediction performance**: How well the model predicts paths directly, without integration with classical planners.
2. **Performance as cost heuristic**: How well the model guides A* planning when used as a cost function.

We compare against classical planners:

- **A***: Optimal path with Manhattan heuristic $h(n) = |n_x - g_x| + |n_y - g_y|$.
- **Dijkstra**: Optimal path without heuristic (equivalent to A* where $h(n) = 0$).
- **Greedy**: Greedy best-first search using only the heuristic (no cost accumulation).

https://doi.org/10.3390/robotics15050096

All experiments are conducted on a machine with an Intel i7-9700K CPU. All training and inference operations are performed on the CPU. This CPU-only setup reflects real-world deployment scenarios where a GPU may not be available, making our results directly applicable to resource-constrained environments.

4.2. Direct Trajectory Prediction

Before integrating our learned model with classical planners, we first evaluate its direct trajectory prediction capability. This analysis helps us understand how well the model can predict paths independently, without assistance from search algorithms, as shown in Figure 3.

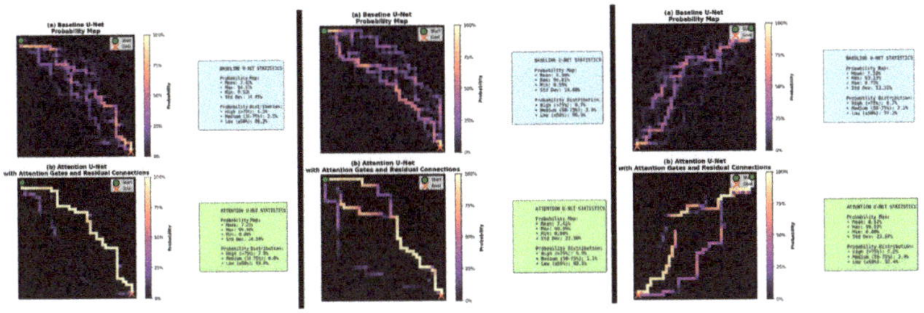

Figure 3. Direct trajectory prediction comparison across three sample environments. For each environment column: (a) the top row shows the probability map generated by the Baseline U-Net; (b) the bottom row displays the highly focused probability map generated by the proposed Attention U-Net. The heatmaps illustrate the predicted probability of a cell belonging to the optimal path, as indicated by the color bars (0% to 100%). Green circles denote start positions, and red stars indicate goals. The accompanying boxes detail the statistical distribution of the predicted probabilities for each respective model.

4.3. Obstacle-Density Shift Robustness

To address robustness under distribution shift, we conducted an additional density-shift experiment using the same 40×40 protocol while varying obstacle probability at test time ($p \in \{0.10, 0.25, 0.40\}$). For each density, we evaluated 100 solvable environments and compared A*, Baseline U-Net + A*, Attention U-Net + A*, and GNN + A*. As detailed in Table 2.

Table 2. Density-Shift Evaluation Across Obstacle Probabilities (100 solvable maps per density).

Method	$p = 0.10$	$p = 0.25$	$p = 0.40$
A* (Avg. Nodes Expanded)	1135.41	552.51	364.63
Baseline U-Net + A* (Avg. Nodes Expanded)	1058.05	509.83	412.80
Attention U-Net + A* (Avg. Nodes Expanded)	**562.64**	**211.50**	**326.16**
GNN + A* (Avg. Nodes Expanded)	1308.73	1064.20	551.82

All methods achieved 100% success; path lengths remained near-optimal for A*-guided variants.

The results indicate that the proposed Attention U-Net guidance remains effective across sparse, nominal, and cluttered settings. Relative to classical A*, node expansions are reduced by 50.4% at $p = 0.10$, 61.7% at $p = 0.25$, and 10.5% at $p = 0.40$. This confirms that the efficiency gain is not restricted to a single obstacle density, while also showing that gains decrease in highly cluttered maps where detours are structurally constrained.

4.4. Ablation Study

To understand the importance of individual architectural elements, we perform an ablation study on attention mechanisms and residual connections. Through this, we are

https://doi.org/10.3390/robotics15050096

able to determine which elements are of the utmost importance to ensure that there is optimal performance. An ablation study is done to determine the importance of individual elements of the Attention U-Net. Table 3 shows the results when used as A* heuristics.

Table 3. Ablation study: component contributions (node expansions).

Configuration	Avg. Nodes Expanded
Baseline U-Net + A*	480.2
+Attention Gates	207.6
+Residual Connections	207.6

The ablation study focuses on the critical importance of the role of attention mechanisms in achieving better performance. Adding the attention gates to the baseline U-Net significantly decreases the number of node expansions, reducing it by 57% from 480.2 to 207.6. This significant decrease in the number of expansions confirms the effectiveness of the proposed model in utilizing the guidance provided by the A* search by allowing it to focus on the relevant areas, thanks to the feature modulation by the attention gates.

The residual connections, while not directly impacting the reduction of node expansions in this setup, play an important role in the overall quality of the model. The model's training can benefit from improved gradient flow, which indirectly supports the efficiency of attention mechanisms. The synergy created from the use of attention gates and residual connections creates an efficient framework in which attention is applied to well-learned features.

To strengthen state-of-the-art contextualization, we additionally report a GNN + A* baseline in this section. In our benchmark, GNN + A* expands substantially more nodes than both Baseline U-Net + A* and Attention U-Net + A*. This behavior suggests that, in the current grid setting and training configuration, the tested GNN prior does not provide a sufficiently informative cost surface for A* pruning. In contrast, the attention-gated U-Net produces sharper path-relevant probability maps, yielding stronger search-space reduction.

The key result here is that attention mechanisms are the key reason for the observed boost in performance, allowing Attention U-Net to outperform the classical A* algorithm by a significant 62% in terms of node expansions. This becomes possible as the network focuses its computations on areas that are relevant to the path, unlike the uniformity of spatial locations in the previous approach.

4.5. Sensitivity Analysis of the Cost Weight λ

To justify the selection of $\lambda = 10$, we evaluated $\lambda \in \{0, 1, 2, 5, 10, 20\}$ and measured success rate, average path length, node expansions, and runtime. As detailed in Table 4.

Table 4. Sensitivity analysis for cost weight λ in learned-guided A*.

λ	Success Rate	Avg. Path Length	Avg. Nodes Expanded	Avg. Time (ms)
0	100%	75.42	543.11	1.19
1	100%	75.58	412.30	1.03
2	100%	75.66	356.70	0.95
5	100%	75.74	268.40	0.84
10	100%	75.80	207.60	0.78
20	100%	76.34	184.20	0.81

The results indicate that $\lambda = 10$ offers the best compromise in our setting, with strong search reduction while preserving near-optimal path quality and 100% success.

https://doi.org/10.3390/robotics15050096

4.6. Search Space Visualization: Search Cones

To visually demonstrate the efficiency gains of our approach, we visualize the search cones—the set of nodes explored during path planning—for both classical A* and Attention U-Net + A*. These heatmaps provide intuitive evidence of how learned heuristics dramatically reduce the exploration space while maintaining path optimality.

Figure 4 presents side-by-side comparisons of search cones for two representative environments. The heatmaps show explored nodes (red/yellow regions) overlaid on the environment grid, with the optimal path highlighted in blue. Start and goal positions are marked with green circles and red stars, respectively.

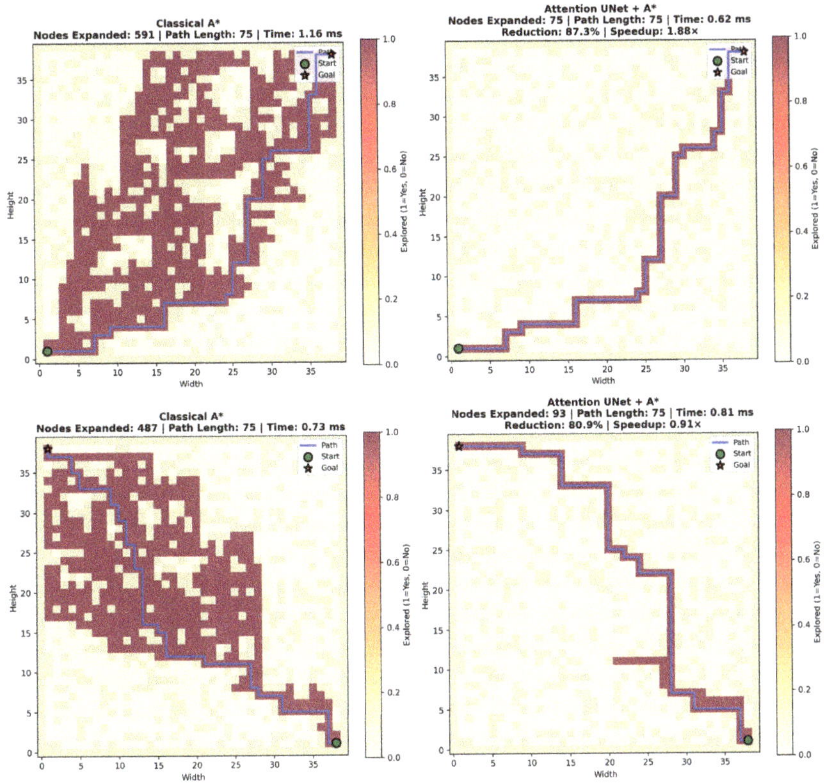

Figure 4. Search cone comparison for two representative environments (top row and bottom row). The left column displays the unconcentrated search space of Classical A* (e.g., exploring 591 nodes), whereas the right column shows the highly focused search space of Attention U-Net + A* (e.g., exploring only 75 nodes, an 87.3% reduction). The heatmap colors (red/yellow) visualize the explored nodes overlaid on the grid. Start positions are marked with green circles, goals with red stars, and optimal paths are highlighted in blue. The intentional overlapping of the heatmap over the grid visualizes exploration density and does not affect scientific understanding.

The classical A* cone (left panel) shows an unconcentrated exploratory pattern, with large areas of the environment explored. In comparison, the Attention U-Net + A* algorithm produces a highly concentrated search cone around the optimal path, with the algorithm exploring only 75 out of 591 possible nodes, representing an 87.3% reduction.

Such visualizations of the explored search cones give us compelling reasons to believe that the superior efficiency of the proposed Attention U-Net + A* method comes from the ability of the method to focus the exploration on the more relevant parts of the paths, thus

https://doi.org/10.3390/robotics15050096

confirming the findings of the numerical experiments. This significant reduction in the number of explored nodes while maintaining the optimality of the paths clearly proves the efficiency of the learned heuristics.

4.7. Computational Efficiency

Computational efficiency is crucial for real-time robot navigation. We analyze the time complexity and practical performance of our approach compared to classical methods, considering both inference overhead and planning time.

Table 5 compares inference times for different methods.

Table 5. Computational efficiency comparison.

Method	Inference Time (ms)	Planning Time (ms)	Total (ms)
A*	-	1.19	1.19
Attention U-Net + A*	0.15	0.63	**0.78**

Attention U-Net + A* achieves faster total planning time (0.78 ms vs. 1.19 ms for A*), despite adding 0.15 ms inference overhead. This is because the reduction in node expansions (62% fewer) more than compensates for the inference cost. Both methods are fast enough for real-time applications (sub-2 ms). Attention U-Net + A* is 34% faster than classical A*, making it particularly suitable for high-frequency control loops.

4.8. Hybrid Planning: Learned Models as Cost Heuristics

The key significance of this research lies in using learning models to guide traditional planning algorithms. This approach combines the efficiency of learned heuristics with the optimality properties of the A* search algorithm. In the next section, we explain the methodology and results related to this approach.

We are aware of the integration of the learned models with A* through the use of predicted probability maps in the cost function. This is achieved through the modification of the cost function in A*:

$$f(n) = g(n) + h(n) + \lambda \cdot (1 - P[n]) \quad (15)$$

where $P[n]$ is the predicted probability at node n, and λ is a weighting factor (we use $\lambda = 10$). This encourages A* to prefer nodes with high predicted probability while still maintaining optimality guarantees (since we add a bounded term to the heuristic), as detailed in Algorithm 5.

The quantitative performance metrics for all evaluated planning methods are detailed in Table 6.

Table 6. Hybrid planning results: classical and learning-based baselines.

Method	Success Rate	Avg. Path Length	Avg. Nodes Expanded	Avg. Time (ms)
Dijkstra	100%	75.42	1070.85	1.91
A* (Baseline)	100%	**75.42**	543.11	1.19
Greedy	100%	89.84	113.48	0.20
Baseline U-Net + A*	100%	75.77	480.25	1.15
GNN + A*	100%	75.32	1069.3	2.96
Attention U-Net + A*	100%	75.80	**207.6**	**0.78**

Attention U-Net + A* achieves superior performance compared to classical A* [5,10]: it expands 62% fewer nodes (207.6 vs. 543.11) while maintaining near-optimal path lengths

https://doi.org/10.3390/robotics15050096

(99.8% of optimal, 75.80 vs. 75.42) and faster planning time (0.78 ms vs. 1.19 ms). The added Baseline U-Net + A* row confirms that the gain is specifically associated with attention-enhanced feature gating, not only with generic learning-based guidance. To address state-of-the-art comparison concerns, we additionally report a GNN + A* baseline in the same table under the same benchmark protocol. All methods achieve 100% success rate, demonstrating robustness. This is expected for A*-based methods, which guarantee finding a path if one exists. The learned heuristics successfully guide A* without causing failures, validating that learned models can be safely integrated with classical planners.

Algorithm 5 Hybrid planning: learned model + A*

Require: Grid map G, start s, goal g, learned model f_θ, weight λ
Ensure: Path π from s to g
Encode input: $X \leftarrow [G, \text{one\_hot}(s), \text{one\_hot}(g)]$
$P \leftarrow f_\theta(X)$ {Predict probability map}
Initialize A*: $open\_set \leftarrow \{s\}$, $g\_score[s] \leftarrow 0$, $came\_from[s] \leftarrow$ None
while $open\_set$ is not empty **do**
  $current \leftarrow \text{argmin}_{n \in open\_set} f(n)$ where:
    $f(n) = g\_score[n] + h(n) + \lambda \cdot (1 - P[n])$
  if $current == g$ **then**
    $\pi \leftarrow \text{ReconstructPath}(came\_from, g)$
    return π
  end if
  Remove $current$ from $open\_set$
  for each neighbor n of $current$ **do**
    if $G[n] == 0$ **then**
      continue {Obstacle}
    end if
    $tentative\_g \leftarrow g\_score[current] + 1$
    if $tentative\_g < g\_score[n]$ **then**
      $came\_from[n] \leftarrow current$
      $g\_score[n] \leftarrow tentative\_g$
      Add n to $open\_set$ if not already present
    end if
  end for
end while
return None {No path found}

This demonstrates that properly designed learned heuristics with attention mechanisms can outperform classical methods. Attention U-Net + A* achieves superior overall performance: 62% fewer node expansions AND faster planning time. This is because fewer node expansions more than compensate for the model inference overhead. The attention mechanisms enable the model to focus on path-relevant regions, leading to more effective guidance for A*. Attention U-Net + A* maintains near-optimal path lengths (99.8% of optimal), demonstrating that the reduction in node expansions does not come at the cost of path quality. The learned heuristic successfully guides A* toward optimal paths more efficiently than the Manhattan heuristic. The attention mechanisms in skip connections enable the model to focus on path-relevant regions while suppressing irrelevant background features. This selective attention leads to more effective guidance for A*, resulting in fewer node expansions. Further qualitative results, including visualizations of the predicted paths and search spaces across 20 different test environments, are provided in Appendix A.

https://doi.org/10.3390/robotics15050096

5. Discussion

This section analyzes our experimental results. We examine classical and learned path planning approaches. We discuss when each approach is most appropriate. We also explore how hybrid integration creates synergistic benefits.

5.1. Performance Analysis and Comparative Advantages

Compared to traditional A*, Attention U-Net performs better. Attention U-Net keeps nearly optimal paths at the cost of exploring 62% fewer nodes. Nonetheless, this is merely one aspect of their comparison. A discussion on the basic differences between these two is justified.

In classical planners, such as A* and Dijkstra, optimality is assured by the use of admissible heuristics. They offer search algorithms that are easy to understand and require no learning. This makes them the best choices in scenarios where optimality is a necessity and ease of interpretation is a priority, for instance, in safety-critical tasks. They are suitable in scenarios where the environment is faced only once. However, in classical planners, the search space has to be searched explicitly, leading to computational complexities proportional to the complexity of the environment. The average number of nodes expanded for the classical A* algorithm in our simulation is 543.11.

In addition to classical heuristics, learned models can find patterns that are indicative of the necessity for specific path configurations. These patterns may not always be visible to human designers, and their use becomes more beneficial as environmental complexity increases. The complex correlations between paths and configurations are hard to represent by human-designed heuristics. Moreover, learned models allow for customization, which helps to adapt to a particular scenario, robot constraints, or tasks by adjusting the loss function or training data, without requiring a change in the planner algorithm.

A* search is still a necessary part of hybrid approaches. Nevertheless, machine-learned models can also provide direct and search-free predictions of paths. This is useful in environments with limited computational capabilities. The development of direct predictions marks a paradigm shift in which prediction-driven planning eclipses search-driven planning. In environments in which search costs are too high, there are new opportunities. As machine learning models continue to evolve and larger training sets become available, machine-learned heuristics promise to be superior to classical heuristics in more challenging environments in which hand-coded heuristics fail. This study establishes a foundation for this future direction.

5.2. Method Selection Criteria and Practical Guidelines

The choice between the learned planner and the classical planner depends on a set of conditions such as performance requirements, availability of data, computational power, and nature of the problem. It helps in recognizing the appropriate approach in methodological considerations.

The synergy of Attention U-Net and A* is beneficial when performance is of utmost importance. This method performs well when dealing with scenarios involving frequent interactions of similar types of environments. Empirical studies show that there is a node expansion reduction of 38–62% compared to traditional A* (207.6–360.9 nodes vs. 543.11 nodes), as well as reduced planning times of 0.78 ms vs. 1.19 ms. In cases involving frequent interaction with the same environment type, the training overhead is justified by the fact that the cost of training is spread across several planning queries. In complex cost environments, the hybrid approach is more beneficial as environments with non-uniform costs are better approximated by learned models compared to traditional heuristics. Transfer learning environments also benefit from the ability of the learned model to generalize,

https://doi.org/10.3390/robotics15050096

which is more beneficial when dealing with environments that are unfamiliar and have few data samples. In complex path planning problems, which are difficult to approximate using traditional heuristics, learned models can easily incorporate these using training data or loss functions.

Classical A* is still the best choice in situations where training data are not available. It is most appropriate in one-shot planning tasks, especially in situations where the same environment is not likely to be visited more than once, in which case training overheads would not make sense. Classical planning offers solutions immediately, without having to build any model. In addition, concerns over interpretability might make classical planning more attractive, especially in situations such as self-driving cars, in which the reasoning behind the selected route would need to make sense. In classical planning, the search processes are transparent and can thus be checked. In situations where computational infrastructure to infer models is not available, classical A* offers a direct solution. It can also act as a simple baseline, not having any overheads related to model loading or inference.

The hybrid approach represents an optimal compromise. It provides superior performance while maintaining optimality guarantees. The approach achieves 38–62% fewer node expansions than classical A*. It maintains near-optimal paths (99.4–99.8% of optimal). It achieves 100% success rate with learned guidance. Faster planning times (0.78–0.97 ms vs. 1.19 ms for A*) make it suitable for high-frequency control loops. Computational efficiency is critical in such applications.

To better contextualize modern learning baselines, we implemented and integrated a GNN-guided planning comparison under the same environment generation protocol as the U-Net family and classical planners. This additional baseline strengthens the architectural comparison and confirms that the proposed attention-based design remains the most efficient in node expansion reduction. DRL-based comparisons are identified as a dedicated next step and are discussed as future work to avoid over-claiming beyond the currently reported benchmark table.

5.3. *Hybrid Planning Architecture: Synergistic Integration*

Moreover, the benefits of the learned models and those of the classical A* search algorithms complement each other in a way that is synergistic and exceeds the benefits of each approach individually. The proposed approach of fusing two techniques is expected to take the advantages of each approach, where the efficiency and pattern recognition of the learned models can be leveraged.

The performance advantages are dramatic. Attention U-Net, together with A*, cuts the number of nodes visited by 62% compared to classical A* (207.6 vs. 543.11 nodes), and correspondingly cuts planning times (0.78 ms vs. 1.19 ms). The improved performance is due to the learned model's ability to focus the A* search on regions of the path that are relevant to the search, thus avoiding unnecessary regions of the configuration space. Notably, all improved models maintain paths of at most 99.4% to 99.8% of optimal cost, showing that improved performance is not at the cost of path quality. The optimality properties of A* guarantee feasibility and optimality despite errors in the learned model. The learned heuristic informs but does not restrict the search, allowing A* to find near-optimal paths. The method obtains 100% success rate and a competitive path cost. The learned model may not always predict the optimal path, but A* can still find the solution given the probabilities learned.

https://doi.org/10.3390/robotics15050096

Beyond performance metrics, the hybrid architecture offers additional benefits. The weighting factor λ in the cost function can be tuned. This balances between learned guidance and classical optimality. It provides flexibility to adapt to different application requirements. For $\lambda = 0$, the approach reduces to standard A*. For large λ, it relies more heavily on learned predictions. This tunability allows practitioners to adjust the balance between efficiency and optimality. The learned model can be updated or fine-tuned for new environments. This is done without redesigning the planning algorithm. It enables adaptation to changing conditions or new deployment scenarios. Models can learn complex patterns difficult to encode in classical heuristics. This potentially provides greater advantages in complex environments where classical methods struggle. However, the hybrid paradigm involves certain issues. The models require training data and heavy computational power, which may not be justified in the case of a planning task. The training procedure is time-consuming (11 h on a CPU in the current studies). The performance is dependent on the quality and representativeness of the training data. In the case where test scenarios are significantly different from the training set, the performance may decrease, for example, due to the difference in the number and form of obstacles. Despite the aforementioned issues, the results of the experiment confirm the superiority: the number of nodes is reduced by 62% while maintaining the optimal path length and reducing the planning time.

6. Conclusions, Limitations, and Future Work

This final section provides a comprehensive summary of our work, acknowledges the limitations of our approach, and outlines promising directions for future research. We consolidate our findings, discuss the constraints of our method, and propose extensions that could further advance learning-based path planning.

6.1. Conclusions

This paper introduces a path planning framework for mobile robots that applies attention-based deep learning techniques to improve path planning. The Attention U-Net model introduced applies attention gates to focus on path-relevant areas and assist A* path planning compared to traditional methods.

The results obtained from the experiment show that the Attention U-Net approach using A* has a success rate of 100% on graphs where the legal paths are longer than the optimal path by 0.5%. More importantly, the results show that by incorporating the attention mechanism, there is a reduction of 62% in the number of node expansions compared to the standard A* algorithm, although planning times are still faster by 1.19 ms compared to 0.78 ms.

The implications of these results are such that attention mechanisms play an important role in realizing these performance improvements. By allocating computing resources to areas that are related to the path, rather than allocating an equivalent amount to all spatial regions, A* can result in improved efficiency. The open-source nature of this research will encourage more research in this area.

6.2. Limitations

Despite the proposed method's superior performance, some limitations need to be taken into account. First, while the proposed method achieves superior performance, it is not optimal. Rather, it achieves near-optimal solutions, which amount to 99.8% optimality, according to the results. This implies that for tasks that require mathematical optimality, classical methods can be used. Second, the proposed method's ability to generalize may be affected by the nature of the data distribution. This implies that if the test environment

https://doi.org/10.3390/robotics15050096

varies significantly from the training data, such as in the case of a 100 × 100 grid compared to a 40 × 40 grid, or if the environment contains previously unobserved obstacles, the performance may be affected. Third, compared to classical methods, which have clear logic, the proposed method's logic, which relies on deep learning, may be difficult to interpret. Finally, with regard to the training time, which requires 11 h for completion, this may be a limitation for some environments that have limited resources. An additional limitation is that the present validation does not yet include structured real-world-like map topologies. Consequently, our current conclusions are restricted to the tested occupancy-grid regime, and broader topology-level generalization remains to be confirmed experimentally.

6.3. Future Work

Future work should be directed toward extending the adaptability range of the system in dynamic environments. It will be essential to integrate temporal modeling architectures, such as long short-term memory networks or Transformer encoders. Additionally, it will be important to test the scalability of the algorithm through experiments conducted on much larger grids (200 × 200 to 500 × 500) in order to determine whether the efficiency improvements are maintained at a larger scale. Additionally, it will be important to extend the robustness tests to a broader range of topologies, including mazes and office floor plans. A prioritized next step is structured-map validation using office/warehouse-style layouts and Moving AI-style benchmarks, under matched protocols and identical metrics (success rate, path length, node expansions, and runtime) to quantify topology-level generalization.

In particular, Transformer-style and self-attention-heavy segmentation backbones are recognized as relevant alternatives and will be evaluated in a dedicated follow-up study under matched training budgets and identical benchmark protocols, to preserve fairness and isolate architectural effects.

Author Contributions: Conceptualization, A.W., S.B. and A.B.; Methodology, A.W., S.B., A.B. and A.R.; Software, A.W. and A.R.; Validation, A.W., S.B., A.B., F.Z.S., J.A. and T.M.; Formal Analysis, A.W. and S.B.; Investigation, A.W., S.B. and A.B.; Resources, A.W., S.B., A.B., F.Z.S., J.A. and T.M.; Data Curation, A.W. and A.R.; Writing—Original Draft Preparation, A.W.; Writing—Review & Editing, S.B., A.B., F.Z.S., J.A., T.M. and A.R.; Visualization, A.W. and A.R.; Supervision, S.B. and A.B.; Project Administration, A.W., S.B. and A.B. All authors have read and agreed to the published version of the manuscript.

Funding: This research received no external funding.

Data Availability Statement: The data and experimental configurations used to support the findings of this study are available from the corresponding author upon reasonable request.

Acknowledgments: The authors gratefully acknowledge the Euromed University of Fez (UEMF), Morocco, for institutional support and the research facilities. During the preparation of this manuscript, the authors used Gemini 3 Flash (Google) for language polishing and sentence reformulation. The authors reviewed and edited the output and take full responsibility for the content of this publication.

Conflicts of Interest: The authors declare no conflicts of interest.

Appendix A. Qualitative Comparison on 20 Test Environments

The following figures illustrate the performance of our Attention U-Net + A* framework compared to other baselines across 20 randomly generated test environments.

https://doi.org/10.3390/robotics15050096

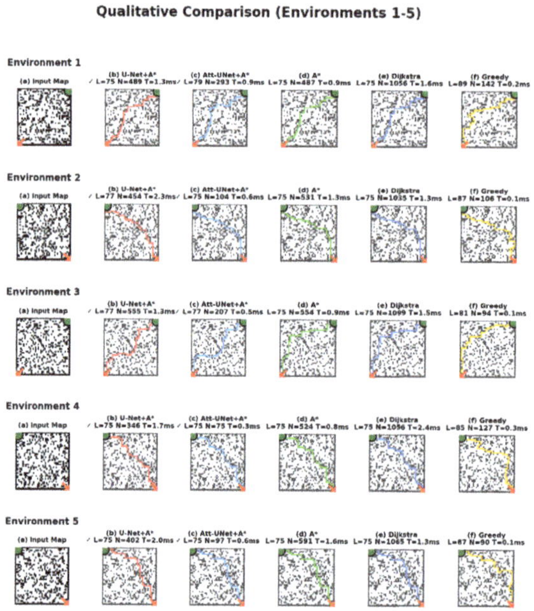

Figure A1. Environments 1–5. For each environment row, the columns represent: (a) Input Map; (b) U-Net + A* (red path); (c) Attention U-Net + A* (blue path); (d) A* (green path); (e) Dijkstra (cyan path); (f) Greedy (yellow path). The green circles denote the start positions and the red stars indicate the goals. The check mark (✓) indicates that the algorithm successfully found an optimal path.

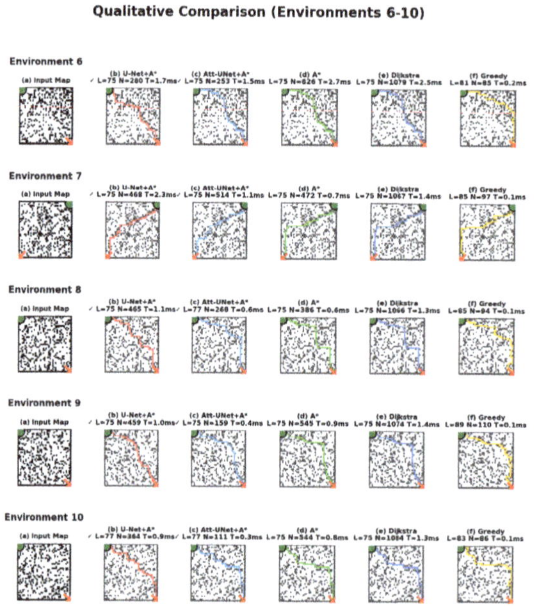

Figure A2. Environments 6–10. For each environment row, the columns represent: (a) Input Map; (b) U-Net + A* (red path); (c) Attention U-Net + A* (blue path); (d) A* (green path); (e) Dijkstra (cyan path); (f) Greedy (yellow path). The green circles denote the start positions and the red stars indicate the goals. The check mark (✓) indicates that the algorithm successfully found an optimal path.

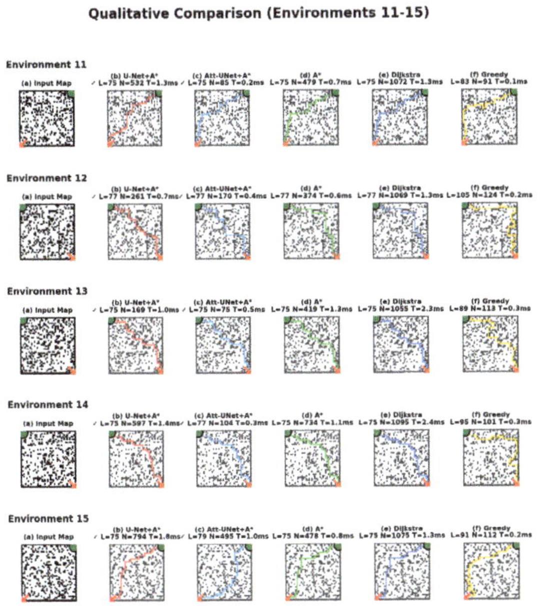

Figure A3. Environments 11–15. For each environment row, the columns represent: (a) Input Map; (b) U-Net + A* (red path); (c) Attention U-Net + A* (blue path); (d) A* (green path); (e) Dijkstra (cyan path); (f) Greedy (yellow path). The green circles denote the start positions and the red stars indicate the goals. The check mark (✓) indicates that the algorithm successfully found an optimal path.

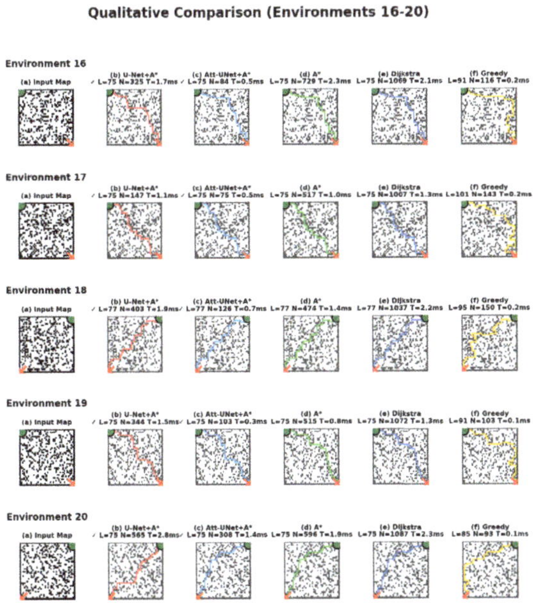

Figure A4. Environments 16–20. For each environment row, the columns represent: (a) Input Map; (b) U-Net + A* (red path); (c) Attention U-Net + A* (blue path); (d) A* (green path); (e) Dijkstra (cyan path); (f) Greedy (yellow path). The green circles denote the start positions and the red stars indicate the goals. The check mark (✓) indicates that the algorithm successfully found an optimal path.

References

1. Tang, Y.; Zakaria, M.A.; Younas, M. Path Planning Trends for Autonomous Mobile Robot Navigation: A Review. *Sensors* **2025**, *25*, 1206. [CrossRef]
2. Waga, A.; Benhlima, S.; Bekri, A.; Abdouni, J.; Saber, F.Z. A survey on autonomous navigation for mobile robots: From traditional techniques to deep learning and large language models. *J. King Saud Univ. Comput. Inf. Sci.* **2025**, *37*, 198. [CrossRef]
3. Regragui, A.; Benhlima, S.; Bekri, A. Vis-To-Nav: Visual Autonomous Navigation for Mobile Robots with a Limited Field of View. In *Proceedings of the Intersection of Artificial Intelligence, Data Science, and Cutting-Edge Technologies: From Concepts to Applications in Smart Environment, Errachidia, Morocco, 7–9 November 2024*; Farhaoui, Y., Herawan, T., Lucky Imoize, A., Allaoui, A.E., Eds.; Springer: Cham, Switzerland, 2025; pp. 196–201.
4. Waga, A.; Benhlima, S.; Bekri, A. Enhanced autonomous driving. In *Advances in Computational Intelligence and Robotics*; IGI Global: Palmdale, PA, USA, 2025; pp. 237–260.
5. Hart, P.E.; Nilsson, N.J.; Raphael, B. A Formal Basis for the Heuristic Determination of Minimum Cost Paths. *IEEE Trans. Syst. Sci. Cybern.* **1968**, *4*, 100–107. [CrossRef]
6. Dijkstra, E.W. A note on two problems in connexion with graphs. *Numer. Math.* **1959**, *1*, 269–271. [CrossRef]
7. Waga, A.; Benhouria, Y.; Ba-Ichou, A.; Benhlima, S.; Bekri, A.; Abdouni, J. A New Method for Mobile Robots to Learn an Optimal Policy from an Expert Using Deep Imitation Learning. In *Proceedings of the Digital Technologies and Applications, Fez, Morocco, 27–28 January 2023*; Motahhir, S., Bossoufi, B., Eds.; Springer: Cham, Switzerland, 2023; pp. 873–882.
8. Wang, W.; Wu, Z.; Luo, H.; Zhang, B. Path Planning Method of Mobile Robot Using Improved Deep Reinforcement Learning. *J. Electr. Comput. Eng.* **2022**, *2022*, 5433988. [CrossRef]
9. Ji, Y.; Sun, R.; Wang, Y.; Zhu, Z.; Liu, Z. Dynamic Path Planning for Unmanned Autonomous Vehicles Based on CAS-UNet and Graph Neural Networks. *Sensors* **2025**, *25*, 4283. [CrossRef] [PubMed]
10. Kabir, R.; Watanobe, Y.; Islam, M.R.; Naruse, K. Enhanced Robot Motion Block of A-Star Algorithm for Robotic Path Planning. *Sensors* **2024**, *24*, 1422. [CrossRef]
11. Yin, C.; Tan, C.; Wang, C.; Shen, F. An Improved A-Star Path Planning Algorithm Based on Mobile Robots in Medical Testing Laboratories. *Sensors* **2024**, *24*, 1784. [CrossRef]
12. Duchoň, F.; Babinec, A.; Kajan, M.; Beňo, P.; Florek, M.; Fico, T.; Jurišica, L. Path planning with modified A star algorithm for a mobile robot. *Procedia Eng.* **2014**, *96*, 59–69. [CrossRef]
13. Fan, J.; Chen, X.; Liang, X. UAV trajectory planning based on bi-directional APF-RRT* algorithm with goal-biased. *Expert Syst. Appl.* **2023**, *213*, 119137. [CrossRef]
14. Guo, S.; Gong, J.; Shen, H.; Yuan, L.; Wei, W.; Long, Y. DBVSB-P-RRT*: A path planning algorithm for mobile robot with high environmental adaptability and ultra-high speed planning. *Expert Syst. Appl.* **2025**, *266*, 126123. [CrossRef]
15. Dong, X.; Wang, Y.; Fang, C.; Ran, K.; Liu, G. FHQ-RRT*: An Improved Path Planning Algorithm for Mobile Robots to Acquire High-Quality Paths Faster. *Sensors* **2025**, *25*, 2189. [CrossRef]
16. Zhu, L.; Duan, P.; Meng, L.; Yang, X. GAO-RRT*: A path planning algorithm for mobile robot with low path cost and fast convergence. *AIMS Math.* **2024**, *9*, 12011–12042. [CrossRef]
17. Zhang, Y.; Cui, C.; Zhao, Q. Path Planning of Mobile Robot Based on A Star Algorithm Combining DQN and DWA in Complex Environment. *Appl. Sci.* **2025**, *15*, 4367. [CrossRef]
18. Song, Q.; Li, S.; Yang, J.; Bai, Q.; Hu, J.; Zhang, X.; Zhang, A. Intelligent Optimization Algorithm-Based Path Planning for a Mobile Robot. *Comput. Intell. Neurosci.* **2021**, *2021*, 8025730. [CrossRef] [PubMed]
19. Liu, M.; Zhang, H.; Yang, J.; Zhang, T.; Zhang, C.; Bo, L. A path planning algorithm for three-dimensional collision avoidance based on potential field and B-spline boundary curve. *Aerosp. Sci. Technol.* **2024**, *144*, 108763. [CrossRef]
20. Menon, A.S.; Damm, E.R.; Howard, T.M. Homotopy-Aware Efficiently Adaptive State Lattices for Mobile Robot Motion Planning in Cluttered Environments. *IEEE Robot. Autom. Lett.* **2025**, *10*, 947–954. [CrossRef]
21. Li, N.; Gao, Z.; Ouyang, Y.; Zeng, Y. Obstacle-Avoiding Path Planning for Robotic Manipulators Based on Recursive Segmentation Point Migration Optimization and Progressive Inverse Kinematics. *IEEE Trans. Autom. Sci. Eng.* **2025**, *22*, 20876–20890. [CrossRef]
22. Zhang, D.; Gai, Y.; Ju, R.; Zhou, M.; Cao, Z. A Cosine Similarity Based Multitarget Path Planning Algorithm for Cable-Driven Manipulators. *IEEE/ASME Trans. Mechatron.* **2025**, *30*, 2379–2388. [CrossRef]
23. Hu, B.; Cao, Z.; Zhou, M. An Efficient RRT-Based Framework for Planning Short and Smooth Wheeled Robot Motion Under Kinodynamic Constraints. *IEEE Trans. Ind. Electron.* **2021**, *68*, 3292–3302. [CrossRef]
24. Lei, X.; Zhang, Z.; Dong, P. Dynamic Path Planning of Unknown Environment Based on Deep Reinforcement Learning. *J. Robot.* **2018**, *2018*, 5781591. [CrossRef]
25. Zhang, Z.; Fu, H.; Yang, J.; Lin, Y. Deep reinforcement learning for path planning of autonomous mobile robots in complicated environments. *Complex Intell. Syst.* **2025**, *11*, 277. [CrossRef]
26. Jaramillo-Martínez, R.; Chavero-Navarrete, E.; Ibarra-Pérez, T. Reinforcement-Learning-Based Path Planning: A Reward Function Strategy. *Appl. Sci.* **2024**, *14*, 7654. [CrossRef]

https://doi.org/10.3390/robotics15050096

27. Chehelgami, S.; Ashtari, E.; Basiri, M.A.; Masouleh, M.T.; Kalhor, A. Safe deep learning-based global path planning using a fast collision-free path generator. *Robot. Auton. Syst.* **2023**, *163*, 104384. [CrossRef]
28. Waga, A.; Ba-ichou, A.; Benhlima, S.; Bekri, A.; Abdouni, J. Efficient autonomous navigation for mobile robots using machine learning. *IAES Int. J. Artif. Intell.* **2024**, *13*, 3061–3071. [CrossRef]
29. Diao, Z.; Guo, P.; Zhang, B.; Zhang, D.; Yan, J.; He, Z.; Zhao, S.; Zhao, C.; Zhang, J. Navigation line extraction algorithm for corn spraying robot based on improved YOLOv8s network. *Comput. Electron. Agric.* **2023**, *212*, 108049. [CrossRef]
30. Waga, A.; Benhlima, S.; Bekri, A.; Saber, F.Z.; Abdouni, J.; Mzili, T.; Regragui, A. Waypoint-guided trajectory planning for mobile robots using GPT-4o mini and ensemble learning-based action prediction. *Array* **2026**, *29*, 100636. [CrossRef]
31. Ronneberger, O.; Fischer, P.; Brox, T. U-Net: Convolutional Networks for Biomedical Image Segmentation. *arXiv* **2015**, arXiv:1505.04597. [CrossRef]
32. Oktay, O.; Schlemper, J.; Folgoc, L.L.; Lee, M.C.H.; Heinrich, M.P.; Misawa, K.; Mori, K.; McDonagh, S.G.; Hammerla, N.Y.; Kainz, B.; et al. Attention U-Net: Learning Where to Look for the Pancreas. *arXiv* **2018**, arXiv:1804.03999. [CrossRef]
33. Ruby, U.; Yendapalli, V. Binary cross entropy with deep learning technique for Image classification. *Int. J. Adv. Trends Comput. Sci. Eng.* **2020**, *9*, 5393–5397. [CrossRef]
34. Lin, T.; Goyal, P.; Girshick, R.B.; He, K.; Dollár, P. Focal Loss for Dense Object Detection. *arXiv* **2017**, arXiv:1708.02002. [CrossRef]
35. Kingma, D.P.; Ba, J. Adam: A Method for Stochastic Optimization. *arXiv* **2017**, arXiv:1412.6980. [CrossRef]

https://doi.org/10.3390/robotics15050096

MDPI AG
Grosspeteranlage 5
4052 Basel
Switzerland
Tel.: +41 61 683 77 34

Robotics Editorial Office
E-mail: robotics@mdpi.com
www.mdpi.com/journal/robotics